T0323714

POWERING THROUGH
THE TRANSITION

POWERING THROUGH THE TRANSITION

Navigating the Energy Sector's Biggest Change Since the Discovery of Oil

MICHAEL DEIGHTON

ELSEVIER

Elsevier
Radarweg 29, PO Box 211, 1000 AE Amsterdam, Netherlands
125 London Wall, London EC2Y 5AS, United Kingdom
50 Hampshire Street, 5th Floor, Cambridge, MA 02139, United States

Notices

Knowledge and best practice in this field are constantly changing. As new research and experience broaden our understanding, changes in research methods, professional practices, or medical treatment may become necessary.

Practitioners and researchers must always rely on their own experience and knowledge in evaluating and using any information, methods, compounds, or experiments described herein. In using such information or methods they should be mindful of their own safety and the safety of others, including parties for whom they have a professional responsibility.

To the fullest extent of the law, neither the Publisher nor the authors, contributors, or editors, assume any liability for any injury and/or damage to persons or property as a matter of products liability, negligence or otherwise, or from any use or operation of any methods, products, instructions, or ideas contained in the material herein.

ISBN: 978-0-323-91754-4

For information on all Elsevier publications
visit our website at https://www.elsevier.com/books-and-journals

Publisher: Megan Ball
Acquisitions Editor: Fran Kennedy-Ellis
Editorial Project Manager: Ellie Barnett
Production Project Manager: Maria Bernard
Cover Designer: Vicky Pearson Esser

Typeset by STRAIVE, India

Working together
to grow libraries in
developing countries

www.elsevier.com • www.bookaid.org

Dedication

For Isabelle and Guy

Contents

Preface

When you've been in the energy sector as long as I have, you start to notice patterns. Every few years, we encounter new challenges. However, I assure you, this time, with the Energy Transition underway, it's different. We're not just shifting gears; we're rebuilding the engine while the car is still speeding down the highway.

Welcome to *Powering through the transition: Navigating the energy sector's biggest change since the discovery of oil*. This book is your guide through the labyrinth transformation from high-carbon to low-carbon energy systems. And let's be clear, it's a labyrinth with no guaranteed exits, only more opportunities for those who dare to navigate it.

I've been in the trenches of the energy sector for over two decades, leading projects and operations for some of the world's top blue-chip organizations. My journey has been anything but linear. From implementing maintenance programs at downstream facilities, leading major complex brownfield projects to spearheading corporate digital innovation initiatives, I've seen firsthand the tidal waves of change that can either drown a company or propel it to new heights.

But enough about me, let's talk about why you should care. The energy transition is not a fleeting trend, it's a seismic shift driven by the urgent need to combat climate change, enhance energy security, and increase the sustainability of our energy systems. If you're an energy manager, engineer, or simply someone with a vested interest in the future of energy, this book will provide you with the tools and techniques you need to thrive in this new landscape.

We'll dive deep into principles of transformational leadership, lean and visual management, agile, and margin improvement methods. These aren't just buzzwords, they are proven strategies that have delivered tangible results for supermajor energy companies. And trust me, they can work for you too. We'll explore smart deployment methods utilizing digital technology and multiskilling of personnel because in this new era, versatility and adaptability are your best friends.

In this book, we'll look at how to visualize development and implementation with flow charts and graphical abstracts. We'll delve into the art of establishing a "Center of Excellence," providing synergy between business groups, supporting the upskilling of the workforce, and driving maximum value out of all contracts. And yes, we'll even make data fun. Because, let's face it, data-driven decision making is not just a nice to have, it's the backbone of any successful energy transition strategy.

One of my favorite parts of this journey has been the chance to speak and participate in the governing body at conferences like ADIPEC, EIC, and GASTECH. It's there that I've seen the spark of innovation in the eyes of fellow professionals, the eagerness to embrace new methods, and the relentless drive to make our industry not just profitable but sustainable. This book captures that spirit, offering practical examples and a glossary of terms to ensure you have everything you need at your fingertips.

So, whether you're in a boardroom strategizing the next big move or out in the field implementing the latest technology, this book is your companion. Together, we'll change the conversation and introduce a new, innovative, and visual approach to navigate the uncertainties in the energy industry. We'll focus on maximizing contract value, improving profitability, and achieving performance excellence.

Let's embrace this journey with a sense of adventure and a dash of humor, because, in the end, the energy transition is not just about reducing carbon footprints; it's about creating a brighter, more sustainable future for all of us.

Michael Deighton, MENG, CENG, MBA, FIMECHE
Senior Vice President and Fellow of the
Institute of Mechanical Engineers

Acknowledgments

First and foremost, I would like to extend my deepest gratitude to my publisher, Elsevier, for having the faith in me to embark upon a second book. Writing another book was a natural progression for me, driven by a desire to delve deeper into the evolving landscape of the energy sector and share more insights that can guide and inspire others during this transformative period. Katie Hamond has supported me on both projects; thank you for your belief in me. Your guidance and support have been invaluable. I am also immensely grateful to my editorial project manager, Ellie Barnett. Ellie, your attention to detail and tireless dedication have been instrumental in bringing this book to life.

To my incredible family, words cannot express how much you mean to me. Isabelle and Guy, you are my world, your laughter, love, and joy are my greatest treasures. Watching both of you grow and flourish has been the most rewarding experience of my life. Your unconditional love and support have been my motivation through every challenge. You both make me strive to be the best version of myself, and for that, I am eternally grateful. My Sister and Brother, I am truly blessed to have such rewarding relationships with you both throughout our lives. Thank you for always being there for me and for being such incredible siblings. To my Mum, your kindness and caring nature are unmatched. You are always putting others before yourself with unwavering love and compassion. Your selflessness, nurturing spirit, and resolute support have given me the confidence to pursue my dreams. You have shown me the importance of empathy, kindness, and resilience. To my Dad, you have been a role model and a source of endless inspiration. Your hard work, dedication, and integrity have shaped who I am today. You have taught me the value of perseverance, the importance of staying true to one's principles, and the significance of leading by example. Your wisdom and guidance have been a constant source of strength and direction throughout my life. I strive every day to live up to the example you have set. I am eternally grateful for the love and support you both have given me, and I am honored to be your son. Your influence and love have made me who I am, and this book is a testament to the incredible impact you have had on my life. Thank you for everything.

To my new and old friends, your friendship and support have been a source of strength, Mick, Paul, Wassim, Patrick, Liam, Hareesh, Simon, Richard, Amale, Nick, and Katya thank you for being there.

Great leaders inspire and drive us to reach new heights. Tush, you have always been there, offering unwavering support and invaluable guidance throughout my career. Your belief in my ability and commitment to my growth have made a profound impact on my professional development, and for that, I am deeply appreciative. I would also like to acknowledge the remarkable leaders who have influenced and supported me over the recent years: Simon N, Simon L, Iain, Tom, Joe, Wayne, John, and Ben. Your strategic vision and ability to navigate complex challenges have been a source of inspiration. Each of you has played a significant role in shaping my professional journey, and I am profoundly grateful for your mentorship and support. Thank you for being exemplary leaders and for leaving an indelible mark on my career.

One of my motivations to embark upon such an ambitious project was through the inspiration of our next generation of students, engineers, and project managers that I have had the pleasure of mentoring and supporting in their early careers. To Scarlett, James, Wasiq, Ming, and Isabelle, you are the future, and I am committed to supporting and guiding you as you embark on your careers.

This book is dedicated to the incredible people who have surrounded me with their love, support, and inspiration. Thank you all for being a part of this journey.

CHAPTER ONE

Introduction

Transformation

The energy sector is currently experiencing a transformative shift, a revolution rivaling the discovery of oil. This unprecedented transformation is propelled by the urgent need to combat climate change, enhance energy security, and increase the sustainability of our energy systems. Renewable energy sources, such as solar, wind, and geothermal, are at the forefront of this transition, gradually replacing fossil fuels.

As we shift toward renewable energy sources, we witness profound changes in energy production, distribution, and consumption. The energy mix is evolving, and there is a growing recognition of the necessity to reduce greenhouse gas emissions, increase energy efficiency, and embrace innovative technologies that foster sustainable energy practices. The digital revolution has also catalyzed a monumental shift in the energy sector, as advanced analytics, machine learning, and artificial intelligence revolutionize industry operations. These innovations empower energy companies to optimize their operations, minimize costs, and enhance their environmental performance.

The climate crisis imposes significant pressure on the oil and gas industry to curb their greenhouse gas emissions and transition toward more sustainable energy sources. Consequently, many companies are diversifying their portfolios and investing in renewable energy sources like wind and solar power. Some are also exploring cutting-edge technologies, such as carbon capture and storage, to diminish their carbon footprint.

It is crucial to understand that this transition to sustainable energy will not happen overnight, and fossil fuels will continue to play a role in the foreseeable future. However, the energy sector is undeniably undergoing a transformative revolution, evolving into a more sustainable, efficient, and technologically advanced industry. New opportunities will emerge for

Powering Through the Transition
https://doi.org/10.1016/B978-0-323-91754-4.00005-4

companies that are ready to adapt, while those that fail to do so, or fail to adapt quickly enough, will face dire consequences.

In this rapidly evolving energy landscape, companies must be agile and proactive in adapting to the pace of change. They must be willing to reinvent themselves, acquire new skills, and forge partnerships with entirely new ecosystems. Successfully navigating this transformation requires companies to embrace innovation and pursue new opportunities. This may involve exploring novel business models, investing in groundbreaking technologies, and adopting fresh approaches to collaboration and partnerships.

To thrive amid this profound transformation, energy companies must be adaptable and innovative. Those who successfully reinvent themselves, cultivate new skills, and establish robust partnerships will be well-positioned to navigate the emerging energy landscape. The ultimate objective is to create a more sustainable, efficient, and technologically advanced industry capable of meeting the energy demands of future generations.

Throughout history, the energy sector has experienced controversy and change, with fluctuations between record profits and drastic cost-cutting measures becoming the norm. These outcomes often stem from global economic influences, supply and demand fluctuations, and extraordinary events that affect oil production or distribution. However, the energy sector is now being reshaped by our collective responsibility to address the climate change crisis. As the world transitions away from fossil fuels and embraces cleaner energy sources, traditional oil and gas companies face the challenge of reinventing themselves to adapt to this changing landscape.

During the 2021 United Nations climate change conference (COP 26), held in Glasgow, Bill Gates, a staunch advocate for green energy and technology investments, warned that some oil giants would ultimately fail, stating that "30 years from now, some of those oil companies will be worth very little." As the world's energy policies shift toward green energy in the fight against climate change, investors are becoming increasingly apprehensive about investing in oil. Demand for green energy will continue to accelerate, further intensified by the commitments made to harness alternative, cleaner energy sources [1].

The magnitude of change required is immense, and the path forward remains largely uncharted for all of us. What is certain, however, is that the energy industry landscape will be unrecognizable as we continue this journey, and there will be numerous casualties along the way. Energy operating companies and service providers alike will be forced to adapt and evolve to survive. An innovative and radical approach is necessary, with a decisive shift toward becoming more efficient, leaner, and focused on radically reducing operating costs and optimizing operational performance. Embracing net-zero emissions is an imperative we must confront with open arms. There is no avoiding this monumental global challenge as we collectively embark on the transition to green energy. As the energy landscape continues to shift, companies that proactively adapt and adopt more sustainable practices will be better positioned to thrive in this new era. The energy sector's future hinges on our ability to innovate, collaborate, and drive forward the development and implementation of cleaner, more efficient, and sustainable energy solutions for generations to come.

How can we take a different approach, to not only survive, but thrive in the new era of the energy sector?

This book is about changing the conversation and responding to the challenge with a fresh new approach. The *Powering Through the Transition* presents an effective, tried, and tested approach that sets a strong foundation to succeed in these uncertain times, to create high-performing teams, drive a culture of innovation and continuous improvement, develop collaborative high-trust client-contractor relationships, and move toward a more dynamic and flexible operating model in the light of the energy transition. Powering through the transition walks energy companies through a process, which examines their current operating models to identify the change required, then to support the change journey, focusing on maximizing the value of contracts by implementing value-add initiatives, lean methodologies coupled with cutting-edge digital solutions to optimize operational performance. It also seeks to implement a framework for continuous improvement to align with the new energy landscape, enhance profitability, and deliver operational excellence. The initiatives described in this book, some of which are well established, are presented in a synergistic framework designed to address the challenges of the energy transition (ET) have been successfully put to the test with major International Energy Companies and Service Providers.

This book is about supporting energy professionals on their change journey, to set them up for success, arming them with the necessary thought leadership and tools to successfully navigate the change. It is important therefore to spend some times to understand the drivers for change in a little more detail so that we can establish a baseline to effectively manage our change journey.

Proven reserves

As of 2021, global oil consumption has remained relatively stable, with the International Energy Agency (IEA), estimating that it mounted to 92.6 million barrels per day (bpd). This represents a slight increase from the previous year, but is still below the prepandemic levels of around 100 million bpd.

In terms of annual consumption per person, the figure has remained around 4.2 barrels per person per annum, although this can vary significantly between countries and regions. The IEA also notes that the COVID-19 pandemic has had a significant impact on oil demand, with lockdowns and travel restrictions leading to a sharp drop in consumption in 2020.

Despite the continued reliance on oil and other fossil fuels, the renewable energy sector has continued to grow rapidly in recent years. In 2020, renewable energy sources accounted for nearly 90% of new power capacity additions globally, according to the IEA. Solar and wind power have seen significant cost reductions, making them increasingly competitive with fossil fuels in many parts of the world.

However, while progress is being made in the transition to clean energy, the challenges ahead remain significant. The IEA estimates that global carbon emissions from energy use are set to increase by 1.5 billion tons in 2021, as the global economy rebounds from the pandemic. This emphasizes the need for continued investment and policy support to accelerate the shift toward a sustainable energy future.

In terms of fossil fuel reserves, it is worth noting that the figures can vary depending on the definition and methodology used. According to the BP Statistical Review of World Energy 2021, global-proven oil reserves stood at 1.7 trillion barrels at the end of 2020. At current consumption levels, this would last for around 47 years. However, this figure does not consider the potential impact of changes in technology, prices, or government policies, which could significantly affect the global demand for, and supply of oil and other fossil fuels [2].

Let us take a moment to process the gravity of this situation.

The world as we know it is on the brink of an energy revolution, a paradigm shift that will require a complete overhaul of the way we consume and produce power. For over a century, we have relied on fossil fuels to power our transportation, homes, and industries. However, this unsustainable dependence on nonrenewable resources has led to the depletion of our planet's finite reserves and an alarming rise in greenhouse gas emissions, contributing to the catastrophic effects of climate change. In just a few short decades, we must transition to a clean and sustainable energy system, one that relies primarily on renewable sources, such as wind, solar, hydro, and geothermal power.

When we talk about proven oil and gas reserves, we need to understand what is being referred to. "Proven reserves" are generally defined as having a 90% or above likelihood of commercial extraction of oil or gas from the reservoir. This contrasts with "probable reserves," which are generally defined as having a much lower probability of recovery of over 50% but under 90%. When we make bold statements, such as "how much oil is left in the world," we are talking about proven reserves. However, the world's proven fossil fuel reserves are just a tiny proportion of the total known fuel resource, and to be considered as a proven reserve, the criteria must be met, in that there must be a high probability (90% or more) of the fossil fuel being extracted profitably. With increasing difficulty to extract oil and gas comes increased cost, and therefore, the Financial Investment Decision (FID) becomes more unlikely.

Paradoxically, with fossil fuels being a limited resource, it would make sense that the reserves would decline every year with their global usage. However, the historical trends suggest that reserves have remained flat over recent years. This is mainly because as more unconventional fuel sources of oil or gas are discovered (e.g., shale gas) and are upgraded to the prestigious "proven" status due to technological advances, they restock the proven reserve supply. However, even with this additional complication to the "proven reserves" equation, proven oil and gas reserves of the elite group of major oil and gas companies referred to as "Big Oil" (ExxonMobil, BP, Shell, Chevron, Total, ConocoPhillips, and Eni) are falling and at an alarming rate. Produced volumes are not being fully replaced with new discoveries. Big Oil lost 15% of its stock levels in the ground in 2020, falling by 13 billion barrels of oil. Their remaining reserves are set to run out in less than 15 years. This is of course unless Big Oil makes new findings.

The task of identifying and extracting oil and gas is becoming more of a challenge because the rate of success is reducing and the investments in exploration are also tending to decline. The declining proven reserves could create major obstacles for Big Oil to maintain stable production levels in coming years without replacing the reserves with new discoveries. In turn, this challenge will have an impact on revenues and pose a major risk in financing energy transition plans. Keeping an eye on the "reserve replacement ratio" is important for oil and gas companies. The reserve replacement ratio is defined as the ratio between new oil that the company discovers through exploration, and the oil it produces. If oil and gas companies want to survive and remain profitable in the long term, they need to maintain a reserve replacement ratio of at least 100% each year. The reserve replacement ratio of Big Oil fell to just 75% in 2015. As a result, it was believed that (in 2016) the world might face an oil shortage of as much as 4.5 million barrels of oil per day by 2035. In 2020, reserve replacement is at a 20-year low and oil companies are replacing just one in six existing barrels with new discoveries [2].

As the complexity of extracting oil increases, so does the cost, and when the cost reaches a point where oil companies cannot extract the oil profitably, the probable reserve becomes uneconomical, and the FID will not be made. Therefore, even if the oil is technically recoverable, this is one reason why any oil reserve estimate must be approached conservatively. The number of oil reserve discoveries is likely to continue to rise each year, with these reserves being categorized as probable reserves. Oil companies are constantly working to make exploration and extraction more reliable and efficient by implementing new technologies and lean methodologies. However, economic recoverability is a different issue, and it depends on oil demand. With the inevitable shift to green energy, oil demand will be increasingly threatened. Although we may have enough oil to last for the next 50 years, the key issue is whether this is enough time to successfully transition to alternative fuel sources before it runs out.

1.5°C

Climate change is a fundamentally important topic. Its impact has affected us all to some degree. Researchers say that it has been around since the first Industrial Revolution, with the first signs of warming from the rise in greenhouse gases as early as 1830 in the tropical oceans and the Arctic. This was nearly 200 years ago, revealing that human-induced climate change was taking place during the mid-19th century [3].

In December 2009, the United Nations Climate Change Conference, known as COP-15, was held in Copenhagen, Denmark. The conference was a crucial opportunity for global leaders and officials to come together to address the urgent issue of climate change and establish a framework for reducing greenhouse gas emissions worldwide.

Despite the high expectations surrounding the conference, the outcome fell short and received widespread criticism. One of the primary criticisms was the absence of a legally binding agreement to reduce greenhouse gas emissions. Instead, the conference produced the nonbinding Copenhagen Accord, which aimed to limit global warming to below 2°C above preindustrial levels.

The conference was also criticized for its lack of inclusion of developing countries. Many developing nations argued that the accord did not adequately address their concerns, such as funding for adaptation measures and technology transfer. The financing of climate change mitigation and adaptation measures was another significant issue at the conference, with many developing countries stating that developed nations had not done enough to provide the necessary funding to help them tackle the effects of climate change.

Furthermore, China and the United States, the two nations responsible for 40% of global carbon emissions, were not willing to offer substantial concessions. Bolivian President Evo Morales stated at the time that "the meeting has failed. It's unfortunate for the planet. The fault lies with the lack of political will by a small group of countries led by the US." In the aftermath of the conference, there was widespread disappointment and frustration with the outcome. Many environmentalists and activists argued that the conference had failed to deliver the necessary action to address the urgent issue of climate change [4].

Despite the criticisms, the conference did have some positive outcomes: It helped to raise awareness of the issue of climate change and highlighted the need for urgent action to reduce greenhouse gas emissions. The conference also led to a formal recognition of the scientific view that the increase in global temperature should be below 2°C. This recognition has since been adopted by many countries and has become a key target in global efforts to address climate change.

Despite such corroborated reservations at COP 15, all 167 countries signed on to the Copenhagen Accord, endorsing the 2°C target. This means that despite the evidence suggesting that 2° may not be enough, all countries have agreed to work toward this target to try and reduce the effects of global warming. While the objective of limiting global warming to 2°C has been widely discussed and adopted in international climate negotiations, recent scientific research suggests that it may not be enough to prevent significant harm to our planet. The effects of climate change have already been felt with the current global temperature increase of only 0.8°C, including the alarming consequences of more acidic oceans, increased flooding, and melting Arctic sea ice.

Is 2° enough...?

Renowned climate experts, Professor Kerry Emanuel and James Hansen, have expressed their concerns about the potential risks associated with a temperature increase of more than 1°C. According to these experts, the current target of limiting global warming to 2°C is inadequate and could lead to long-term disaster for our planet.

Professor Emanuel argues that "any number much above one degree Celsius involves a gamble... and the odds become less and less favorable as the temperature goes up." He emphasizes that a temperature increase of more than 1°C could result in severe consequences for the planet, including rising sea levels, increased frequency and intensity of natural disasters, and significant damage to ecosystems and human communities. Similarly, James Hansen, the top climatologist at NASA, cautions that the 2° target is a "prescription for long-term disaster." He argues that the current trajectory of global greenhouse gas emissions is on track to exceed this target, and urgent action is needed to avoid catastrophic consequences.

The warnings from these experts highlight the urgent need for aggressive action to mitigate the effects of climate change and protect the planet for future generations. This includes reducing greenhouse gas emissions through the adoption of cleaner and more sustainable energy sources, implementing policies to promote energy efficiency, and investing in climate adaptation measures to prepare for the impacts of global warming [5].

Tipping points

There is a strong consensus in the scientific community that human emissions are driving global warming and climate change and that some scientists believe "tipping points" have already been breached or are imminent. The seriousness of tipping points and the need for urgent action to limit global warming and reduce emissions is a major problem for the future of our planet, as small perturbations can trigger cascading effects that may have widespread and long-lasting consequences.

But what exactly are tipping points and why are they important?

Tipping points refer to conditions in the climate system where a small change in one part of the system can become self-sustaining and lead to irreversible impacts with severe consequences for humanity. These points of no return are a growing concern for scientists, policymakers, and the public.

Climate tipping points occur when changes in significant parts of the climate system, known as "tipping elements," become self-sustaining beyond a certain threshold. If triggered, these points can lead to catastrophic consequences, such as rising sea levels due to melting ice sheets, the collapse of biodiversity hotspots like the Amazon rainforest, and the release of carbon from thawing permafrost.

Several reports on climate tipping points have all come to the same alarming conclusion: even a global warming of 1°C (which we have already surpassed), puts us at risk of triggering these tipping points. This highlights the critical importance of limiting additional warming as much as possible.

Lenton et al. published a study, which aimed to identify the potential climate tipping elements that could trigger abrupt and irreversible changes in the Earth's climate system. The study highlighted nine potential tipping elements, which are key components of the Earth system that could experience abrupt and irreversible change once a tipping point is reached. These tipping elements included Arctic sea ice, Greenland ice sheet, West Antarctic ice sheet, Atlantic thermohaline circulation, El Nino-Southern Oscillation, Amazon rainforest, Boreal forest, ocean methane hydrates, and permafrost.

The authors defined tipping points as thresholds beyond which a small change in forcing leads to a large, potentially irreversible, response of the climate system. They identified the tipping points for each potential tipping element and assessed the timescales and impacts of their potential tipping. For example, the Arctic sea ice tipping point was defined as the point at which summer sea ice extent falls below 10%–20% of its 1979–2000 average extent, while the Atlantic thermohaline circulation tipping point was defined as the point at which this circulation collapses, leading to a significant cooling of the North Atlantic region.

The study found that some of the tipping elements could potentially reach their tipping point even if global warming is limited to 2°C above preindustrial levels, the target set by the Paris Agreement. The authors emphasized the need for immediate and significant reduction in greenhouse gas emissions to prevent the crossing of tipping points and avoid potentially catastrophic and irreversible impacts on the Earth's climate system.

The study also highlighted the potential feedback mechanisms between tipping elements, where the crossing of one tipping point could trigger the crossing of another, potentially leading to a cascade of tipping events. For example, the melting of the Greenland ice sheet could lead to changes in ocean circulation, which could trigger the collapse of the Atlantic thermohaline circulation, further exacerbating global climate change.

Lenton concluded that if the global average temperature rises by 2°C above preindustrial levels, it could trigger abrupt and irreversible changes in the Earth's climate system, leading to catastrophic consequences for the planet's ecosystems and human societies. They argued that this temperature threshold should serve as a critical policy target for mitigating the worst impacts of climate change [6].

In October 2021, the 26th United Nations Climate Change Conference of the Parties (COP-26) was held in Glasgow, UK. The conference was considered a crucial turning point for the global community's fight against climate change, given the urgency of the situation. It was attended by world leaders, policymakers, scientists, and activists who came together to discuss the most pressing environmental issues of our time.

One of the most significant outcomes was a new number: 1.5°C. The pact aims to limit global warming to 1.5°C, a goal that scientists now believe is essential to prevent the most severe consequences of climate change. The pact is a significant achievement as it signals the commitment of the global community to take concrete action to address the climate crisis. To achieve the goal of limiting global warming to 1.5°C, significant efforts are needed to reduce CO_2 emissions and implement methods to remove CO_2 from the atmosphere. This is a challenging task that requires the world and the energy sector to adapt and transition to sustainable energy sources. The transition will require significant investments in renewable energy, energy-efficient technology, and carbon capture and storage methods to reduce carbon emissions. Achieving this goal will not be easy, and it will require a collective effort from governments, industries, energy companies, and individuals to make the necessary changes.

The effects of climate change are already visible in many parts of the world, including more frequent and severe weather events, sea-level rise, and habitat destruction. Limiting global warming to 1.5°C is crucial to prevent the most severe consequences of climate change and protect the planet and future generations. The transition to sustainable energy sources is an urgent task that must be addressed now to mitigate the devastating effects of climate change [7]. This presents a significant challenge for the world and the energy sector.

The energy transition

The term "energy transition" somehow sounds like it is a well-oiled machine that when activated, will smoothly change from one reality to another. In fact, it is far more complicated. Throughout history, energy transitions have been difficult, and this one is even more challenging than any energy transition in the past. As the name implies, an energy transition is a structural change in an energy system. This phrase carries a lot of emphasis on the transition part, because this is, in general, a radical shift from the predominant current energy system to a new one. Historically, there have already been numerous energy transitions.

In Daniel Yergin's book, "A New Map," he pegs the beginning of the first energy transition to January 1709, when an English metalworker named Abraham Darby figured out that he could make better iron by using coal rather than wood for heat. The 19th century is known as the "century of coal" but not until the beginning of the 20th century did coal overtake wood as the world's number 1 energy source. The first transition was protracted to say the least [6].

Past energy transitions have actually not been "transitions" at all, and they have rather been "energy additions," where one source overlays another. For example, oil was discovered in 1859, and it did not surpass coal as the world's primary energy source until a century later in the 1960s. Conversely, the world uses almost three times as much coal as it did in the 1960s today.

Until the mid-19th century, traditional biomass (the burning of solid fuels, such as wood or crop waste) was the dominant source of energy used across the world. With the Industrial Revolution came the staggering rise of coal. By the turn of the 20th century, around half of the world's energy came from coal and half still came from biomass. Throughout the 1900s, the world adopted a broader range of sources. First oil, gas, then hydropower. It was not until the 1960s that nuclear energy was added to the mix. Renewables, such as solar and wind, were only added much later during the 1980s.

What stands out from this 200-year history of global energy use is that energy transitions have been very slow in the past. It has taken roughly a century for a particular energy source to become dominant. While this is true of the past, there are signs that this is changing. Some recent energy transitions happened very quickly. In the UK, for example, nearly two-thirds of electricity came from coal power in 1990. By 2010, this had fallen to just below one-third, and in the decade that followed it fell to around 1% [6].

Today, when we think about energy sources, a diverse mix comes to mind—coal, oil, gas, nuclear, hydropower, solar, wind, and biofuels. However, the diversity of our current energy system is a relatively recent development. If we were to journey back a few centuries, we would find that humanity relied on merely one or two primary energy sources. When examining the historical data on global primary energy consumption, we can trace it back to as early as the year 1800.

This historical perspective highlights the evolution and diversification of our energy sources over the past couple of centuries. The changes in energy consumption patterns reflect the advancements in technology, economic development, and societal needs that have occurred over this period. As we continue to seek cleaner and more sustainable energy sources to address climate change and environmental concerns, it is essential to recognize the progress we have made and the opportunities that lie ahead in further diversifying our energy mix. This ongoing evolution will play a critical role in shaping the future of energy consumption and the global transition to sustainable energy systems [8].

Global primary energy consumption by source

Primary energy is calculated based on the 'substitution method' which takes account of the inefficiencies in fossil fuel production by converting non-fossil energy into the energy inputs required if they had the same conversion losses as fossil fuels.

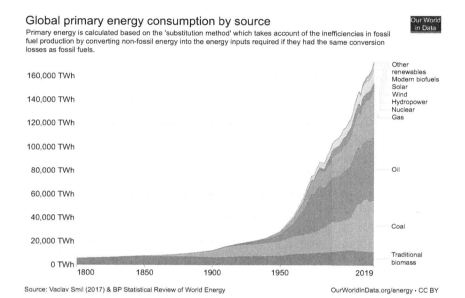

Source: Vaclav Smil (2017) & BP Statistical Review of World Energy   OurWorldInData.org/energy · CC BY

During the 1800s, there was a significant shift in energy sources, as society transitioned from wood to coal. This was followed by another major transformation in the 1900s, when coal was gradually replaced by oil and gas. These historical energy transitions were driven by factors, such as cost and efficiency, and led to changes in energy technologies and services, including heating, cooling, lighting, and mechanical power. Unlike these earlier transitions, which were primarily motivated by economic factors, the impending energy transition is fundamentally different.

The upcoming energy transition is not just an addition to existing energy sources; it is anticipated to be a complete overhaul of the energy foundation that supports today's $86 trillion global economy, which relies on hydrocarbons for 80 percent of its energy needs. The goal is to establish a net-carbon-free energy system, with carbon capture technology, to power an estimated $185 trillion economy by 2050. Achieving this massive transformation in less than 30 years—much of which needs to be accomplished within the next nine years—is an immense challenge.

The complexity of this energy transition becomes evident when considering the extent to which the world depends on oil and gas. It is not just about replacing gasoline-powered vehicles with electric ones, which, by the way, are composed of roughly 20 percent plastic. Plastics are derived from petrochemicals, which are produced using oil and natural gas as raw materials. These plastics can be found in a wide array of products that we use daily. For instance, we consider the packaging of food and beverages. From plastic bottles and containers to cling wrap and food storage bags, plastics play a crucial role in preserving and protecting our food. Similarly, in our homes, everyday items like toothbrushes, disposable razors, and plastic kitchenware are all made from oil and gas derivatives.

Electronics are another area where oil and gas products are prevalent. The casings for smartphones, laptops, and flatscreen are all made from plastic materials, while wiring insulation and circuit boards also utilize plastic components. Additionally, many appliances and devices contain lubricants derived from oil, ensuring their smooth and efficient operation.

In the world of fashion, synthetic fabrics, such as polyester, nylon, and acrylic, are made using petrochemicals. These materials are commonly used in clothing, footwear, and accessories, demonstrating yet another way oil and gas permeate our everyday lives. In agriculture, oil and gas derivatives play crucial roles in the production of fertilizers, pesticides, and even the fuel that powers farm machinery. As the world transitions away from fossil fuels, alternative sources of energy and raw materials will be required for agricultural production to meet growing global food demands.

Paradoxically, plastics are integral components of wind towers, solar panels, and even the lubrication of wind turbines. For instance, wind turbine blades are constructed with composite materials that include plastic resins, while solar panels require plastic polymers for encapsulation, providing protection and structural support. The transition to renewable energy sources will necessitate finding alternative materials or solutions for these components. With the ubiquitous presence of oil and gas, it is clear how deeply entrenched our dependence is throughout various aspects of modern life, emphasizing the challenges and complexities involved in achieving a comprehensive energy transition.

Although renewable energy has made great strides in recent years, it still only accounts for a small percentage of the world's energy supply. According to the IEA, renewables made up just 29% of global electricity production in 2020, with most of the remaining share coming from fossil fuels. In the article "The Energy Transition Confronts Reality" by Yergin, the energy shift is faced with various challenges, some of which have been dismissed or undervalued. The global energy infrastructure is extensive and intricate, having been developed over numerous decades. It encompasses everything from oil and gas wells, refineries, and pipelines to power stations, transmission lines, and distribution networks. To replace this entire system with a clean and renewable energy infrastructure necessitates substantial time, effort, and financial resources [9].

The 2015 Paris climate conference set the objective of carbon neutrality, while COP26 in Glasgow made significant progress on the methods for achieving that goal. Nevertheless, we may still have much to discover about the complexities that await us in the energy transition.

It is projected that the genuine cost of the worldwide energy shift will be around $15 trillion in the next 30 years. Renewables will attract the majority of this investment. While this means the energy transition is far from inexpensive, no credible source ever claimed that going green would be cheap. However, the investments allocated to expanding wind, solar, and related systems will not be the sole expenses incurred during the transition. There may also be considerable environmental costs.

Implementing measures

The global energy landscape is rapidly evolving, and the shift toward renewable energy sources, coupled with the rapid advancements in digital technologies, is creating a seismic shift in the oil and gas industry. The magnitude of change associated with the Energy Transition lends itself to era period of profound change. Companies are being forced to reevaluate their traditional business models and adapt to new ways of working to remain competitive and relevant in this rapidly changing landscape.

To stay ahead of the curve, oil and gas companies must recognize the magnitude of the changes required and commit to reinventing themselves in multiple ways. These companies are facing mounting pressure to reduce costs and increase profitability, so they must look for ways to optimize their operations and streamline their processes, be leaner and more efficient. Digital solutions, such as automation and machine learning, are becoming a norm, which can drive greater efficiency and productivity while reducing operational costs, a topic which will be explored in detail in Chapter 4.

In Chapter 5, we will present how data-driven approaches have become increasingly important in the energy sector, more importantly now than ever with the energy transition. As companies seek to make informed decisions, they must collect, analyze, and interpret data to gain insights into their operations. These data can be used to improve operational efficiency, reduce downtime, and optimize decision-making processes.

Investing in data analytics platforms is essential for companies that are looking to reinvent themselves in the energy transition era and harness the power of data. These platforms enable companies to collect and process vast amounts of data, providing them with real-time insights into their operations. By using advanced analytics tools, companies can identify patterns and trends in their data, which can help them to optimize their operations and reduce costs.

In addition to investing in data analytics platforms, companies must also develop teams of skilled digital engineers to analyze and interpret data. These digital engineers play a critical role in turning data into actionable insights that can drive better decision-making. They are responsible for developing data models, analyzing data sets, and interpreting the results to provide actionable recommendations. For example, data-driven approaches can have a significant impact in reducing downtime of an asset. By collecting and analyzing data from various sources, companies can identify patterns that may lead to equipment failure or downtime. By addressing these issues proactively, companies can reduce downtime, improve asset utilization, and increase productivity. Data-driven approaches can also enable companies to develop more accurate and reliable forecasting models, which can help to plan and allocate resources more effectively. This can be especially important in the context of the energy transition, where demand patterns and market conditions may be changing rapidly.

The adoption of more collaborative contracting models is now also a key shift in the new energy transition era. To drive innovation and achieve shared goals and maximize the value of contracts, companies need to work closely with clients, suppliers, and partners. This has become especially prominent because of the need to navigate the uncertainty associated with the changing energy landscape and may involve the deployment of more unconventional business models, such as alliances, partnership, and outcome-based contracts, which incentivize suppliers to deliver results rather than providing products or services for the lowest cost. Embracing collaboration enables companies to foster innovation and leverage the expertise of their partners and suppliers to drive growth.

Given the huge amount of change in the industry, it is imperative for companies to gear themselves up to be flexible enough to respond rapidly. They must make continuous improvement and innovation as an essential part of their business strategies. This involves accepting new technologies and exploring novel approaches, while maintaining a mindset of ongoing learning and development. Companies should prioritize research and innovation to stay ahead of their competitors and be able to react swiftly to market changes. By embracing a culture of continuous improvement and innovation, businesses can remain flexible and adaptable in response to the ever-changing energy sector. In the following chapters, we will cover the critical change goals that are necessary to navigate the energy transition era.

References

[1] CNBC Article, Bill Gates Predicts Oil Companies will be Worth Very Little in 30 Years—Here's Why, 2021.

[2] Oil and Gas Journal/Rystad Energy, Oil Production Set to Rise Over Run in Less than a Decade, 2021.

[3] Global Power | Research + Innovation, Chris Luekemann, The Global Energy Outlook, 2023.

[4] The Guardian Article, Rich and poor countries blame each other for the failure of Copenhagen deal, 2009.

[5] Kent, Joint Program on the Science and Policy of Global Change. Massachusetts Institute of Technology Climate and Economic Forecasting, 2021.

[6] D.I. Armstrong McKay, A. Staal, J. Abrams, R. Winkelmann, B. Sakschewski, S. Loriani, I. Fetzer, S. Cornell, J. Rocha, Exceeding 1.5°C Global Warming Could Trigger Multiple Climate Tipping Points, 2022.

[7] COP 26—UN Climate Change Conference, Main Outcomes of the COP26, 2021.

[8] H. Ritchie, How the World's Energy Use has Changed Over the Last Two Centuries, 2023.

[9] D. Yergin, The Energy Transition Confronts Reality, 2023.

CHAPTER TWO

Reinvention—From oil giants to pioneers of sustainable energy

Beyond fossil fuels

In the face of a rapidly changing global energy landscape, oil and gas companies find themselves confronted with the colossal task of transforming from fossil fuel giants into pioneers of sustainable energy solutions. This transformation is fueled by many factors including, increasing environmental concerns, regulatory shifts, and the relentless advent and adoption of new technologies and digital solutions. These unassailable International Oil Companies (IOCs) are now faced with an obligatory change in their operational frameworks. As the world moves toward embracing cleaner, more eco-friendly energy sources, the pressure to adapt is crucial not only for their survival but also for the long-term sustainability of our planet. This transformation, though laden with uncertainty, offers an unprecedented chance to reinvent and redefine the role of these industry leaders in a new era of energy production, consumption, and stewardship.

With the global population projected to grow by two billion over the next quarter of a century, the demand for energy will rise at an equally alarming rate. Satisfying this need will necessitate the deployment of all available energy sources both traditional and alternative. While alternative energy sources continue to make significant strides, they alone will not suffice to meet the world energy demands. It is a reality that we all need to come to terms with that hydrocarbon-based energy sources will remain essential in fulfilling most the world's energy requirements and form part of a balanced energy system for the foreseeable future.

To successfully navigate this transition, oil and gas companies must face a multifaceted change journey, which requires a complete reinvention of their business models. Traditional IOCs have relied heavily on vertical integration, controlling every aspect of the oil production process from exploration to refining and distribution. However, to earn the prestigious title

Powering Through the Transition
https://doi.org/10.1016/B978-0-323-91754-4.00008-X

"International Energy Companies" (IECs), these companies must explore and adopt a different approach that encompasses a balanced portfolio of green and carbon-based energy, collaboration, innovation, and agility and embraces the digital revolution. Some IECs are partnering with technology firms and renewable energy companies to develop cutting-edge solutions for energy storage, distribution, and management. Others are establishing innovation hubs or incubators to foster the development of new technologies and business models that can drive growth in a low-carbon economy. New long-term cross-sector partnerships are being formed to balance the risk stepping into this new landscape. By forging partnerships with renewable energy and technology players, the new age of IECs can help drive innovation, scale, and cost-competitiveness in the sustainable energy sector.

Big Oil has led the way in reassessing their business models and strategies. As oil and gas giants, they have traditionally thrived on the extraction and distribution of fossil fuels, but now recognize the need to adapt to the evolving energy landscape. Consequently, these once fossil-fuel-loving giants began a transformative journey, reinventing themselves to not only survive but thrive in this brave new world of global energy transition.

Mounting pressures

Traditional oil and gas companies and service providers are facing a magnitude of pressures as a result of the changing landscape, with the inevitable shift to green energy supply, increased competition from new renewable energy and technology players, new more stringent environmental legislation, and the rapid development and adoption of new technology. As countries around the world implement policies to reduce their dependence on fossil fuels, the long-term outlook for oil consumption appears increasingly uncertain. This decline in demand threatens the profitability and viability of IOCs, forcing them to reevaluate their business models and investment strategies.

Another challenge facing IOCs is the rapid growth of renewable energy sources, such as solar and wind. As the cost of these technologies continues to fall, renewables have become increasingly competitive with traditional fossil fuels. This shift in the energy market has led to a surge in investment in renewables, with many governments and businesses seeking to capitalize on the benefits of clean, sustainable energy. Consequently, IOCs are grappling with the need to adapt and diversify their portfolios to stay relevant in the changing energy landscape. In addition to the rise of renewable energy,

IOCs are also contending with growing environmental regulations aimed at curbing greenhouse gas emissions and mitigating the impacts of climate change. Governments across the globe are implementing policies, such as carbon pricing, emissions trading schemes, and stricter fuel efficiency standards, which have a direct impact on the operations and profitability of IOCs. These regulations not only increase the cost of doing business for oil companies but also heighten the reputational risks associated with continued investment in fossil fuels.

The advent of new technologies, such as electric vehicles (EVs), advanced batteries, and smart grids, is further challenging the dominance of IOCs in the energy sector. These innovations are helping to facilitate the transition to a low-carbon economy by enabling more efficient use of energy resources and reducing the reliance on fossil fuels. As the pace of technological change accelerates, IOCs must embrace innovation and find ways to take advantage and integrate these new technologies into their operations or risk being left behind.

Reinvention and rebrand

In the wake of the energy transition, energy companies are increasingly recognizing the need to reinvent themselves. A major part of this process is predicated on public acceptance, and as such, IOCs are redefining their corporate strategies and rebranding themselves. This shift involves investing in renewable energy sources as well as adopting new business models and broadening their focus to encompass a more diversified and sustainable approach to energy production. By rebranding themselves as IECs, they aim to demonstrate their commitment to the energy transition and position themselves as leaders in the global move toward a low-carbon economy. One notable example of this rebranding effort is the Norwegian company Equinor, which changed its name from Statoil in 2018 to reflect its evolving focus on clean energy solutions. This name change represented a strategic shift away from being a predominantly oil-based company toward becoming a broader energy company with a strong focus on renewables, especially offshore wind power. Equinor has since set ambitious targets for expanding its renewable energy capacity and reducing its carbon emissions.

Shell has been gradually increasing its investments in renewable energy and low-carbon technologies, with a focus on wind, solar, and hydrogen. In 2016, the company established a dedicated division called "New Energies" to oversee its investments in renewable energy projects and new technologies. Shell has also set an ambitious target to become a net-zero emissions company by 2050.

BP has been redefining its corporate strategy and setting ambitious goals to become a net-zero company by 2050. To achieve this, BP has outlined plans to increase its investments in low-carbon energy sources and scale back its oil and gas production. In addition to investing in renewable energy projects, BP has also launched a new business division called BP Launchpad, which focuses on investing in and scaling innovative energy startups that align with its low-carbon goals.

ExxonMobil has recently shown signs of increased interest in clean energy and has been investing in carbon capture and storage (CCS) technologies, algae-based biofuels, and hydrogen fuel cells. In 2020, ExxonMobil created a new business unit called ExxonMobil low-carbon solutions, focusing on commercializing low-carbon technologies and reducing emissions.

These rebranding efforts by IOCs to position themselves as IECs signal a recognition of the need for a more diversified and sustainable approach to energy production. By broadening their focus, investing in renewable energy sources, and adopting new business models, these companies aim to stay competitive in a rapidly changing market and contribute to the global transition to a low-carbon energy future.

Renewable energy and low-carbon technologies

As the energy transition gains momentum, IOCs are increasingly investing in renewable energy to diversify their portfolios and to future proof their businesses. Recognizing the growing demand for clean energy and the need to reduce greenhouse gas emissions, IOCs are actively exploring opportunities in wind, solar, and other renewable energy projects. Additionally, they are investing in EV infrastructure and energy storage technologies to further support the shift toward a low-carbon economy.

Several IOCs have made significant investments in wind energy projects. For instance, Equinor has been at the forefront of offshore wind development with projects, such as Hywind, the world's first floating wind farm, and the Dogger Bank Wind Farm, set to become one of the world's largest offshore wind farms upon completion. Similarly, Shell has also entered the offshore wind market with investments in projects like the Borssele III/IV wind farm in the Netherlands and the Mayflower Wind project in the United States. In the solar energy sector, IOCs are also making huge strides. TotalEnergies has been expanding its solar portfolio through acquisitions and partnerships. It acquired a majority stake in SunPower, an American solar energy company, and has since been involved in numerous solar projects around the world, such as the Al Dhafra Solar PV plant in the United Arab Emirates, set to be one of the world's largest solar power plants.

IOCs are not limiting their investments to wind and solar energy; they are also exploring other forms of renewable energy, such as bioenergy and geothermal power and investing heavily in low-carbon technologies to support the global shift toward a more sustainable and low-carbon energy future. These efforts include research and investment in CCS, hydrogen production, and biofuels—technologies that have the potential to significantly reduce greenhouse gas emissions and contribute to the overall energy transition. These efforts play a critical role in helping IOCs evolve into IECs, better equipped to face the challenges posed by the changing energy landscape and contribute to global efforts to reduce greenhouse gas emissions.

CCS: A crucial pillar of the energy transition

In the vast landscape of the energy transition, one technology has emerged as a crucial pillar in the pursuit of a low-carbon future: CCS. As the urgency to mitigate greenhouse gas emissions grows, CCS has garnered significant attention for its potential to reduce carbon dioxide (CO_2) emissions from large-scale industrial sources and play a vital role in achieving ambitious climate targets. Carbon capture and underground storage (CCUS) encompasses a suite of technologies designed to capture CO_2 emissions from industrial processes, such as refining, power generation, and steel manufacturing, before they are released into the atmosphere. Once captured, the CO_2 is transported and stored deep underground in geological formations, preventing its release and subsequent contribution to global warming.

The significance of CCS lies in its ability to address emissions from sectors that are challenging to decarbonize through other means. Industries, such as cement and steel production, rely on high-temperature processes that generate substantial CO_2 emissions inherently linked to their production methods. By capturing and storing these emissions, CCS offers a viable solution to tackle the carbon footprint of these industries, bridging the gap between their current emissions and a sustainable, low-carbon future.

Numerous energy companies and organizations around the world have recognized the potential of CCS and are investing heavily in this technology. Their commitment to developing and deploying CCS solutions highlights its importance in the energy transition. One prominent example is

Shell, which has made significant strides in CCS development and implementation. Through its Quest project in Alberta, Canada, Shell has successfully captured and stored over 5 million metric tons of CO_2 since 2015, demonstrating the feasibility and scalability of CCS on an industrial scale. Shell is also actively involved in other CCS initiatives globally, emphasizing its commitment to advancing this technology.

Another notable player is Equinor has long been at the forefront of CCS efforts. Equinor's Sleipner field, located in the North Sea, has been storing CO_2 since 1996, preventing millions of metric tons of emissions from entering the atmosphere. Equinor continues to expand its CCS portfolio, including participation in projects like Northern Lights, a collaboration to develop transport and storage infrastructure for captured CO_2 in Norway.

The commitment of these energy companies, among others, highlights the growing recognition of CCS as a critical technology in the energy transition. Their investments and initiatives demonstrate a collective effort to accelerate the development and deployment of CCS solutions on a global scale.

Beyond the energy sector, governments and international organizations are also recognizing the importance of CCS in achieving climate goals. Supportive policies, regulatory frameworks, and financial incentives are being established to facilitate the deployment of CCS projects and encourage collaboration among stakeholders.

As the urgency to address climate change intensifies, CCS stands as a vital tool in the energy transition, offering a pathway to significantly reduce CO_2 emissions from industrial sources. Through the dedication and investments of energy companies, as well as the support of governments and organizations, CCS is poised to play a transformative role in the pursuit of a sustainable, low-carbon future.

The rise of hydrogen

In the vast landscape of the energy transition, one fuel has been making waves as a promising contender to reshape our energy future: hydrogen. While many perceive hydrogen as a new addition to the energy mix, it is a fuel that has quietly existed for many years, playing an integral role in the refining process and now emerging as a prominent fuel of choice in the new energy transition era.

For decades, hydrogen has been an unsung hero within the refining industry. Its significance lies in its ability to facilitate vital processes that enable the production of cleaner and more efficient fuels. In the refining process, hydrogen is used in a method called hydroprocessing. This process involves the removal of impurities, such as sulfur, from crude oil and other hydrocarbon feedstocks. During hydroprocessing, hydrogen is introduced to the feedstocks in the presence of catalysts, leading to chemical reactions that result in the removal of undesirable elements. This purification process is critical for producing fuels that comply with environmental regulations and improve their performance. Hydrogen's affinity for bonding with impurities and its role as a reducing agent make it an invaluable component in this refining process.

However, despite hydrogen's integral role in refining, it has often been overshadowed by other fuels when it comes to broader energy discussions. This has led to a common misconception that hydrogen is a newfound solution in the energy transition. In reality, hydrogen has been silently working behind the scenes, refining fuels and contributing to cleaner energy sources for years.

As the world grapples with the urgent need to combat climate change and transition to a low-carbon economy, companies across various sectors are recognizing the potential of hydrogen as a versatile, sustainable, and clean fuel. In response, these forward-thinking companies are stepping up their efforts to produce hydrogen on a larger scale, creating the necessary infrastructure, and investing in research and development (R&D) to unlock its full potential.

The new energy transition era has prompted a paradigm shift, where hydrogen is no longer limited to its refining applications. It is now being harnessed as a direct fuel, especially in sectors where electrification or battery technologies face challenges. Hydrogen's remarkable attributes, such as high energy density, fast refueling, and long-range capabilities, make it a particularly suitable fuel for heavy-duty transportation, shipping, aviation, and industrial processes.

Moreover, hydrogen's potential as an energy carrier and storage medium has gained significant attention. Excess renewable energy generated from sources like wind and solar can be used to produce hydrogen through electrolysis, where water is split into hydrogen and oxygen using electricity. The hydrogen produced can then be stored and utilized when renewable energy generation is low or when there is a peak in energy demand. This capacity to store and release energy on-demand makes hydrogen a valuable tool in balancing the intermittent nature of renewable sources, enhancing the stability and reliability of the grid.

Recognizing the transformative potential of hydrogen, governments, policymakers, and international organizations are increasingly supporting its development through favorable policies, regulations, and financial incentives. The growing momentum behind hydrogen in the energy transition is accelerating the deployment of hydrogen infrastructure, research into efficient production methods, and the advancement of fuel cell technologies.

Companies in the energy transition era are embracing hydrogen as a vital component of their low-carbon strategies. They are investing in cutting-edge technologies to produce hydrogen from renewable sources, such as wind and solar, thereby reducing its carbon footprint and maximizing its sustainability. Leveraging hydrogen as a fuel enables these companies to contribute to a greener and more sustainable future, where carbon emissions are significantly reduced, and the impacts of climate change are mitigated.

As hydrogen takes center stage in the energy transition, it is essential to recognize its longstanding presence and its critical role in the refining process. IOCs are actively investing in hydrogen production technologies, particularly green hydrogen produced using renewable energy. For example, Repsol has partnered with Iberdrola, a Spanish utility company, to develop a green hydrogen project for industrial use. Similarly, Shell is involved in several hydrogen projects, such as the NortH2 project in the Netherlands, which aims to produce green hydrogen using offshore wind power.

TotalEnergies: A case study on reinventing for the energy transition

TotalEnergies is one of the world's leading oil and gas companies. Like many other IOCs, it has faced the challenge of adapting to the global energy transition by redefining its strategy and operations. Embracing this change, the company rebranded itself as TotalEnergies to reflect its commitment to becoming a broad-energy company. One of the most interesting and impactful aspects of its transformation has been the emphasis on operational efficiency and digitization, enabling the company to optimize its existing operations while positioning itself at the forefront of innovative energy solutions (Fig. 2.1).

TotalEnergies has invested heavily in digitizing its operations to enhance efficiency, reduce costs, and minimize the environmental impact of its operations. The company has implemented advanced analytics, artificial intelligence (AI), and machine learning (ML) technologies to optimize its oil and

Fig. 2.1 IECs.

gas production, processing, and distribution operations. These digital solutions have led to significant improvements in asset performance and efficiency, reduced downtime, and data-driven solutions for better-informed decision-making.

A prime example of commitment of TotalEnergies to operational efficiency and digitization is its "Refinery of the Future" project at the Donges refinery in France [1]. The company has introduced several digital solutions, including real-time data monitoring and analysis for process optimization and energy efficiency improvements, and predictive maintenance technologies, which use ML algorithms to analyze equipment performance data and predict potential failures before they occur and advanced process control systems, which leverage AI to optimize refining processes, resulting in reduced energy consumption and emissions. These innovative solutions have led to a significant reduction in energy consumption and greenhouse gas emissions, demonstrating TotalEnergies' dedication to sustainability and efficiency.

TotalEnergies has also embraced Virtual Reality (VR) technology to optimize its operations and enhance safety. The company has deployed VR solutions for remote monitoring and control of offshore platforms, reducing the need for on-site personnel and minimizing operational risks. This digitization effort has not only improved operational efficiency but has also contributed to a safer working environment for employees.

Recognizing that the energy transition requires collaborative efforts, TotalEnergies has established partnerships with various technology companies, research institutions, and startups. The company actively participates in open innovation initiatives, such as the Open AI Energy Initiative (OAI), which aims to develop AI-driven solutions to address the energy sector's challenges. The OAI is a collaborative effort aimed at leveraging AI to tackle the complex challenges facing the energy sector. Bringing together industry leaders, technology companies, research institutions, and startups, the initiative focuses on developing AI driven solutions that enhance efficiency, sustainability, and resilience in energy operations. Through OAI, participants work on innovative projects ranging from optimizing energy production and distribution to reducing carbon emissions and improving predictive maintenance.

TotalEnergies' emphasis on operational efficiency and digitization has been a crucial component of its transformation from an oil and gas giant to a broad-energy company. The importance of embracing digital transformation and operational efficiency cannot be understated as key drivers for success in the rapidly changing energy landscape.

New relationships

The energy transition has had a significant impact on the relationships between energy companies. This period of rapid change and uncertainty has necessitated increased collaboration and interdependence among industry players, as well as greater engagement with other industries, contractors, and stakeholders. As a result, traditional relationships have been evolving to adapt to the new energy landscape.

One of the most striking changes in the energy sector has been the blurring of boundaries between traditional oil and gas companies, utilities, and renewable energy firms. As IOCs diversify their portfolios and invest in renewable energy sources, they are increasingly partnering with, or even acquiring, renewable energy companies. This trend is fostering new relationships and collaborations, as IOCs learn from the expertise of these firms and jointly develop projects in areas, such as wind, solar, and energy storage.

In addition to fostering collaboration within the energy sector, the energy transition has also led to increased engagement between energy companies and other industries, particularly in the development and deployment of new technologies. For example, advancements in EVs and their associated infrastructure have created opportunities for collaboration between IOCs, automotive manufacturers, and technology companies. Similarly, the growth of digitalization and the Internet of Things (IoT) in the energy sector has resulted in partnerships between energy firms and technology companies to develop innovative solutions for smart grids, energy management, and emissions monitoring.

The energy transition has also highlighted the need for contractors and service providers to work more closely with their clients in the energy sector. As projects become more complex and technologies evolve, contractors must adapt to new requirements, regulations, and standards. Collaborative partnerships with clients can help ensure that contractors understand the specific challenges and opportunities associated with the energy transition, allowing them to tailor their services and solutions accordingly, a topic that we will explore further in Chapter 3. Furthermore, this collaboration can help to mitigate risks and uncertainties, leading to more successful projects and long-term relationships.

In this new era of energy transition, the relationships between energy companies and their stakeholders are also evolving. Companies must engage more actively with regulators, investors, and local communities, ensuring that their sustainability goals and commitments are aligned with broader societal expectations. This involves increased transparency and proactive communication, as well as greater involvement in policy discussions and the development of industry-wide standards and best practices.

The uncertainties and complexities associated with the energy transition necessitate a more collaborative and adaptive approach to relationships within and beyond the energy sector. By fostering partnerships and alliances, energy companies can navigate the challenges of this new landscape more effectively, pooling resources, sharing knowledge, and driving innovation to achieve a more sustainable energy future. Ultimately, the energy transition represents an opportunity for the industry to redefine itself and forge stronger, more resilient connections that will enable it to thrive in the face of change.

A good example of how relationships between IOCs are changing due to the energy transition can be observed in the formation of the Oil and Gas Climate Initiative (OGCI). Launched in 2014, the OGCI is a voluntary CEO-led initiative that brings together leading IOCs, including BP, Chevron, Eni, ExxonMobil, Occidental, Petrobras, Repsol, Shell, and TotalEnergies, with the aim of accelerating the industry's response to climate change. The initiative foster collaboration and knowledge-sharing among its member companies, enabling them to jointly invest in low-carbon technologies, develop innovative solutions, and share best practices in areas, such as CCUS, methane emissions reduction, and energy efficiency. By pooling their resources and expertise, these traditionally competitive IOCs are working together to drive progress toward a more sustainable and low-carbon energy future.

In addition to facilitating collaboration within the oil and gas industry, the OGCI also engages with external stakeholders, including governments, nongovernmental organizations, financial institutions, and other industries. This engagement enables the initiative to align its efforts with broader societal goals and contribute to the development of climate-related policies, standards, and best practices. The OGCI exemplifies how the energy transition has prompted IOCs to rethink their traditional relationships and adopt a more collaborative approach to address the complex challenges posed by climate change. Working together and engaging with a diverse group of stakeholders enables these companies to accelerate their transition to low-carbon energy sources and better position themselves for success in the evolving energy landscape.

Enter the new players

In recent years, the oil and gas sector has undergone major change, fueled by investors' growing concerns about the ability of traditional oil and gas companies to yield satisfactory returns in the future. This apprehension stems mainly from the rapidly shifting dynamics within the energy sector and the emergence of new technologies and alternative energy sources.

The accelerating energy transition has generated a demand for an integrated vendor and operator ecosystem capable of effectively driving returns. This new era of the energy sector requires collaboration among various stakeholders in the sector, including producers, service providers, suppliers, consumers, and regulators. Historically, and even paradoxically, traditional oil and gas companies have been reluctant to adopt new technologies and digital solutions. This resistance, however, to innovation is now threatening their long-term survival. In fact, according to Deloitte's Digital Maturity Index Study, which will be discussed in Chapter 4, the oil and gas industry is generally a technology laggard when compared to other industries. This lag in technological adoption has paved the way for innovative and disruptive players to enter the market (Fig. 2.2).

New entrants, including tech giants, such as Google, Apple, and Tesla, have quickly embraced digital technologies to reshape the industry and challenge existing business models. By incorporating digital technologies, such as IoT, AI, and big data analytics, these companies have been able to create a more efficient, cost-effective, and sustainable energy ecosystem.

Google and Apple have both made significant strides in the energy sector in recent years, taking on traditional oil and gas companies like Shell and ExxonMobil and taking market share. One example of this is Google's entry

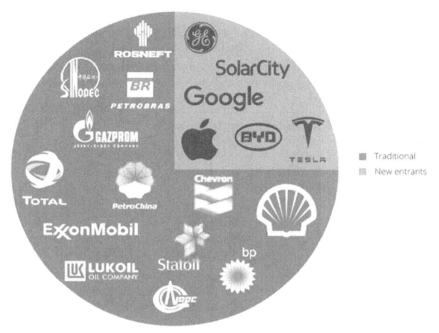

Fig. 2.2 Enter the new players.

into the renewable energy market through its investments in wind and solar energy. Google has invested in a number of large-scale renewable energy projects, including wind farms and solar power plants, with the goal of powering its operations with 100% renewable energy. Similarly, Apple has also made significant investments in renewable energy, investing in large-scale renewable energy projects, including wind and solar, and has also launched its own solar energy subsidiary. Investing in renewable energy and armed with their digital tool kit and innovation and disruptive mindset, both Google and Apple have been able to compete with traditional energy companies in the renewable energy market and take market share.

New service providers are also entering the energy sector, competing with traditional oil and gas service contractors. These new service providers are often focused on digital technologies and analytics, offering services like predictive maintenance, advanced analytics, and remote monitoring. One example of this is GE Digital, a company that provides software and analytics solutions for the energy sector. GE Digital's software allows energy companies to optimize their operations, reduce downtime, and improve efficiency, competing with traditional oil and gas service contractors in the process.

The new players disrupting the Energy sector are another alarm bell about the need to transform in the wake of the energy transition. As the energy sector continues to evolve, traditional energy companies and service contractors must continue to innovate, embrace digital solutions and new technology, and adapt in order to stay relevant and competitive.

Preparing for the change journey

The energy transition has set energy companies on a transformative journey, unparalleled in their history. To be successful in this changing landscape, these companies are embarking on a profound reinvention of themselves. This "change journey" encompasses a multitude of simultaneous shifts, including the digital revolution and the advent of AI. The digital revolution brings forth an era of connectivity, automation, and data-driven insights, enabling energy companies to optimize operations, enhance efficiency, and provide personalized services. Simultaneously, the rise of AI presents opportunities for predictive analytics, intelligent decision-making, and the development of smart energy systems. Moreover, the changing energy landscape demands a shift toward renewable energy sources, decarbonization efforts, and sustainable practices, with a new age of energy company now competing for market share. Energy companies are reimagining their business models, diversifying their portfolios, and embracing innovation to become key players in the clean energy transition. This change journey is a challenging but necessary path, as companies strive to secure their future by aligning with the imperatives of environmental sustainability, technological advancement, and evolving customer expectations.

It is a given that IOC companies are diversifying their energy portfolios by incorporating renewable and low-carbon technologies in line with the energy transition. They are investing in wind, solar, and other clean energy sources and decarbonization of their assets, striving to reduce their environmental impact and align with global climate goals. This involves optimizing operations for greater energy efficiency, investing in CCS technologies to mitigate greenhouse gas emissions, minimizing flaring and methane leaks among many other initiatives. Additionally, some IOCs are considering the divestment of high-emission assets, such as coal-fired power plants and oil sands operations, to accelerate their transition to low-carbon energy sources. These decarbonization efforts are critical for a successful transition to IEC territory. However, as the energy landscape evolves, oil and gas companies must buckle up and embark on a transformative journey to adapt to the global energy transition and consider a host of other aspects along the way. This pivotal transition requires a paradigm shift in how these companies operate, strategize, and innovate. To ensure a smooth ride, there are several critical elements that need to be considered as they gear up for the journey. These critical elements are introduced in this chapter and will be expanded on in the subsequent chapters of this book.

A new age of leadership

There is an imperative for a new kind of leadership capable to embrace the rate of change and navigate the rapidly transforming, complex energy landscape. This new age of leaders needs to possess a unique set of traits and capabilities to effectively navigate the challenges and respond to the rate of change.

The Energy Transition (ET) era requires visionary thinking. Leaders need to have a clear and compelling vision for their organizations' role in the energy transition. They must prioritize sustainability and consider the environmental, social, and economic impacts of their decisions. They should be committed to decarbonization, energy efficiency, and responsible resource management and set ambitious sustainability targets and outlining a strategic roadmap to achieve these goals. Visionary leaders must be forward-thinking, recognizing emerging trends and opportunities, and positioning their organizations to capitalize on them.

New leaders must be adaptive and resilient. The energy transition is marked by uncertainty and rapid change, making adaptability and resilience essential leadership traits. Leaders must be willing to adjust their strategies and business models in response to evolving market dynamics and be prepared to pivot when necessary. This includes being open to embracing new technologies, exploring alternative revenue streams, and reevaluating traditional business practices.

The energy landscape is becoming increasingly technology-driven, with advancements in renewable energy, energy storage, and digitalization of operations. A successful energy transition relies heavily on the development and adoption of new technologies. New leaders must possess technological literacy and an innovation and technology-driven mindset. Leaders need to foster a culture of innovation within their organizations, encouraging experimentation, risk-taking, and continuous learning. This may involve investing in R&D, establishing partnerships with technology firms, or creating innovation hubs to incubate new ideas and solutions.

Although collaboration and partnerships are nothing new in the energy sector, the complex nature of the energy transition demands a more collaborative approach, as no single organization or sector can address all the challenges independently. Effective leaders should be adept at building strategic partnerships and alliances with other industry players, governments, research institutions, and even competitors. This collaborative mindset can facilitate knowledge-sharing, resource pooling, and the development of new, shared solutions.

There must be a strong emphasis on talent development. As the energy landscape evolves, the skillsets required within the industry will also change. Leaders must prioritize talent development, ensuring that their organizations are equipped with the necessary expertise to adapt and thrive in the new energy era. This includes investing in workforce training, promoting diversity and inclusion, and attracting top talent from various backgrounds and disciplines.

Chapter 10 will explore the new age of leaders that are needed to take the helm of energy companies, and they are required to be adaptable, visionary, possess technological literacy, and have a focus on sustainability, resilient, emotionally intelligent, and a commitment to continuous learning. These traits are important for new leaders to embrace the rate of change that is happening and navigate the rapidly transforming energy landscape.

Collaboration and strong partnerships

Building strong partnerships with stakeholders, including governments, communities, suppliers, and customers, is essential to ensure a collaborative and coordinated approach to the energy transition. As the energy transition gains momentum, energy companies are increasingly turning to digital and innovative technologies to supercharge their operations by enhancing efficiency, optimize operations, and reduce environmental impact. To achieve these objectives, they are actively engaging in strategic partnerships, joint ventures, and open innovation initiatives with technology providers, startups, and research institutions. These collaborations allow energy companies to access cutting-edge technologies, leverage external expertise, and foster innovation within their organizations.

In line with the change journey, energy companies are forming strategic alliances with technology providers and startups to develop and deploy new digital solutions in areas, such as data analytics, AI, automation, and the IoT. For example, BP has partnered with AI company Beyond Limits, which marks a significant step toward leveraging cutting-edge technology to enhance the efficiency and effectiveness of oil and gas exploration and production. BP's partnership aims to develop advanced analytics and ML algorithms specifically tailored for the energy sector. According to BP, these tools will have the potential to revolutionize the decision-making process, optimize operations, and improve overall performance in the oil and gas sector. The partnership demonstrates BP's commitment to embracing digital innovation and underscores the industry's recognition of the transformative power of AI. As stated in a BP press release, the collaboration with Beyond Limits aims to "combine human knowledge with machine learning and AI capabilities to deliver step-change improvements in our performance and create a more sustainable and resilient future." The collaboration signifies an exciting development in the ongoing evolution of the energy sector, where traditional practices are being augmented by advanced technologies to drive progress and address complex challenges [2].

Similarly, Shell has teamed up with Microsoft to accelerate its digital transformation journey. Through this partnership, Shell aims to drive innovation, optimize operations, and enhance its overall business performance by leveraging Microsoft's cloud platform and its suite of AI and IoT tools. Shell's intention is to improve decision-making processes and streamline operations across its global operations. The collaboration between Shell and Microsoft highlights the growing recognition among energy companies of the immense potential of digital technologies to revolutionize the industry and drive sustainable growth [3].

Energy companies are also entering into joint ventures with technology companies to co-develop and commercialize innovative solutions. For instance, Chevron has formed a joint venture with carbon capture technology provider Svante to develop and deploy large-scale, low-cost carbon capture solutions for industrial applications. These joint ventures enable energy companies to share the risks and costs associated with developing new technologies while benefiting from the technical expertise of their partners.

To tap into the broader innovation ecosystem, energy companies are launching open innovation programs, such as corporate venture capital funds, incubators, and accelerators, which provide funding and support to startups and researchers working on breakthrough technologies. For example, TotalEnergies operates a venture capital arm, "TotalEnergies Ventures," which plays a crucial role in driving innovation and supporting the development of startups in key sectors of the energy sector. With a focus on renewable energy, energy storage, and digital technologies,

TotalEnergies Ventures actively seeks opportunities to invest in promising startups that are shaping the future of the energy landscape. By providing funding and expertise, it enables these startups to accelerate their growth, scale their technologies, and contribute to the energy transition [4].

Energy companies are also forging partnerships with universities and research institutions to access cutting-edge R&D capabilities. These collaborations often involve joint R&D projects, technology licensing agreements, or the establishment of dedicated research centers focused on areas, such as advanced materials, energy efficiency, and emissions reduction. For example, ExxonMobil has partnered with the Global Climate and Energy Project at Stanford University to research and develop low-emission energy technologies, including advanced biofuels and carbon capture solutions.

Engaging in these various forms of collaboration is key for energy companies to access external expertise, technologies, and innovation capabilities to support their change journeys and navigate the energy transition, drive efficiency improvements, and remain competitive in the rapidly evolving energy landscape. These partnerships also highlight the growing recognition among energy companies of the importance of embracing digital transformation and innovation as key enablers of their transition. We shall delve in to the topic of collaborative relationships and contracting in Chapter 3.

Operational efficiency

Energy companies are compelled to prioritize operational efficiency to maintain their competitiveness and adaptability in the evolving energy landscape. The global shift toward a low-carbon future has led to increased competition due to the blurring of energy market as technology players enter the market, coupled with more stringent environmental regulations. This has put more pressure on companies to improve their operational efficiency and reduce operating costs. Furthermore, the rapid advancements in digital technologies, such as AI, big data analytics, and IoT, naturally impact the oil and gas industry's appetite to invest and adopt these innovations. These companies are adopting a range of strategies and technologies to streamline their processes, enhance decision-making, and minimize expenses.

Operational efficiency intuitively involves exploring and adopting digital technologies. These technologies can help enhance decision-making, improve asset management, and reduce downtime, thus leading to more efficient operations. As the cost of digital solutions and technology has become more affordable in recent years, investing in new technologies to support operational efficiency improvements is also becoming increasingly beneficial.

Harnessing new digital solutions and technologies, such as AI, ML, and automation, enables energy companies to optimize their operations, automate workflows, and achieve significant productivity gains. The adoption of digitization is a critical aspect of this transformative process. Digital tools aid in streamlining processes, enabling real-time data analysis, and refining decision-making. By integrating digital technologies into their core operations, companies can unlock major efficiency gains and reduce operating costs.

ExxonMobil has been actively investing in digital technologies as part of their energy transition strategy to boost operational efficiency, leveraging advanced analytics, ML, and AI. One example is the implementation of advanced predictive maintenance program for their refineries. This program utilizes advanced analytics and ML algorithms to monitor equipment performance and predict potential failures. Identifying potential issues before they escalate allows ExxonMobil to reduce unplanned downtime, enhance equipment reliability, and achieve significant cost savings. We will explore the digital revolution and how digital solutions can be applied to optimize operational performance in Chapter 4.

Data-driven decision-making

The energy sector has seen a significant shift toward data-driven strategies in recent years as the benefits have become more apparent and cost of implementing such strategies significantly reduced. Energy companies are increasingly deploying data-driven strategies to improve their operations, make more informed decisions, and become leaner and more efficient.

Data-driven approaches enabling energy companies to make informed decisions, optimize operations, and drive innovation. By harnessing the power of data analytics and insights, energy companies can gain a deeper understanding of their operations, identify patterns, and uncover hidden opportunities for efficiency improvements. These approaches can enhance asset management, predictive maintenance, and supply chain optimization, leading to cost savings and improved reliability. Furthermore, data-driven approaches enable energy companies to optimize their energy generation and consumption, supporting sustainability goals and reducing environmental impact. Leveraging data to drive decision-making and innovation enables energy companies to stay ahead of the curve, navigate complex challenges, and unlock new avenues for growth in the rapidly transforming energy landscape.

Service providers are also deploying data-driven strategies to better serve their customers and stay competitive in the rapidly evolving market. They are increasingly adopting data-driven approaches to assess performance, capture, analyze, and act on data with confidence. Leveraging advanced analytics, ML, and AI, service providers can gain valuable insights into various aspects of their operations, such as contract or projects performance, workforce efficiency, and equipment performance. They can also leverage the data to support business development activities, such as identify new trends and opportunities in the market, develop more targeted services and solutions, and improve their overall efficiency.

The sheer volume of data being generated by energy companies today means that traditional methods of data analysis are no longer sufficient. The importance of data-driven approaches in the energy sector has become an imperative that cannot be ignored. Instead, companies need to leverage advanced analytics tools and ML algorithms to gain insights from their data and drive operational improvements. Chapter 5 looks at what it means to be data-driven and how to go about making this transition.

Energy transition workforce

As we have discussed, it is no understatement that the energy sector is undergoing a profound transformation, driven by changes in regulations, market dynamics, and the widespread adoption of digital solutions and technologies. Consequently, there is a central need to develop the workforce. However, this endeavor is far from simple. The process of workforce development entails overcoming various challenges and complexities.

One of the key aspects is the redefinition of job roles and introduction of new ones within the industry. As digital technologies and automation become increasingly prevalent, certain traditional job roles may become obsolete, while new roles emerge. For instance, with the rise of renewable energy and smart grids, there is a growing need for experts in energy storage, grid optimization, and data analysis. Moreover, as sustainability takes center stage, there is a demand for professionals specializing in environmental impact assessment, clean energy project management, and carbon accounting. New job roles are now also starting to ramp up because of the digital revolution, for example, in AI, ML, and digital applications as companies are seeing the benefit of these to their businesses. Redefining job roles involves identifying emerging skill sets, aligning them with evolving industry needs, and equipping employees with the necessary knowledge and expertise.

Another challenge in workforce development lies in bridging the skills gap. As the industry embraces digitalization, there is a shortage of workers with the required technical competencies. Upskilling and reskilling programs are essential to address this gap. Companies need to invest in training initiatives that provide employees with the knowledge and skills to leverage digital technologies effectively. This may involve partnerships with educational institutions, the implementation of in-house training programs, or the utilization of online learning platforms. Equipping the workforce with the necessary digital literacy and technical skills is important so that energy companies can enhance operational efficiency, drive innovation, and remain competitive.

Attracting and retaining top talent poses another challenge in workforce development. The energy sector competes with other sectors, such as technology and finance, for individuals with expertise in digital technologies and sustainability. To overcome this, companies need to cultivate an employer brand that appeals to these professionals. This entails offering competitive salaries, providing opportunities for career advancement, fostering a culture of innovation, and promoting work-life balance. Workforce transformation is a multifaceted undertaking that requires redefining job roles, bridging the skills gap, and attracting top talent. Job roles must be aligned with changing industry needs, training is needed to acquire new skills, and an attractive work environment needs to be created to build a resilient and competent workforce. It requires a cultural shift for the organization to appreciate that in this new ET era, innovation and experimentation are needed. A culture where employees feel empowered to try new things and take risks to drive innovation and stay ahead of the competition is required. We will explore how this can be achieved in Chapter 9.

Center of excellence (CoE)

As the tangible effects of climate change intensify, nations, industries, and individuals increasingly commit to reducing their carbon emissions and moving toward cleaner, more sustainable energy sources. This shift represents both a significant challenge and a unique opportunity for energy companies. Navigating this intricate and dynamic landscape demands a strong commitment to innovation, data-driven decision-making, and digital transformation. Energy companies are thus faced with the dual task of capitalizing on these changes not only to ensure their survival but also to redefine their roles in a rapidly evolving market. The introduction of CoEs within these companies plays a crucial role in this transition. CoEs act as specialized units designed to foster innovation, facilitate knowledge-sharing, and standardize best practices across the organization, enabling these companies to manage the complexities of integrating new technologies and approaches effectively.

These CoEs are vital in orchestrating the transition of energy companies toward more sustainable practices. They serve as hubs of expertise, pooling cross-functional teams that bring together specialized knowledge and skills essential for adopting cutting-edge technologies and fostering innovation. The role of CoEs can be likened to the principle of "sharpening the saw" from Stephen Covey's influential work, "The 7 Habits of Highly Effective People," which advocates for continuous self-improvement and care. In a similar vein, CoEs help energy companies to continuously enhance their capabilities and adapt to the changing energy landscape, ensuring they remain competitive and aligned with environmental goals. Chapter 6 will delve into the theory and practical applications of CoEs in energy companies, exploring their strategic importance and their pivotal role in driving the energy transition. Through this exploration, it will become clear that CoEs are not merely functional units but are crucial to the strategic reorientation of energy companies in the face of the global shift toward renewable energy and sustainability.

Project delivery assurance

This energy transition characterized by the increased adoption of renewable energy sources, technological advancements, and evolving regulatory frameworks offered both opportunity and challenge. Central to successfully managing this transition is the sector's capacity to execute projects that are not only innovative but also resilient to the uncertainties inherent in this shift. The dynamic nature of the energy transition, with frequently evolving project scopes and still-developing technologies, underscores the necessity for a robust, structured, yet flexible approach to project delivery. This flexibility is essential, allowing energy companies to adapt to changing conditions, integrate new technologies, and tackle unforeseen challenges while maintaining the integrity of project outcomes.

"Delivery Assurance" therefore emerges as a critical component within the project management frameworks of the energy sector. It involves a suite of practices and principles aimed at ensuring projects are completed on time, within budget, and in line with both the strategic objectives of the organization and the broader goals of the energy transition. This chapter delves into the critical nature of a structured yet agile approach to project delivery, exploring how it supports the success of the energy transition.

Reinvention

Energy companies are embarking upon a transformational journey to reinvent themselves the face of the to a rapidly evolving energy landscape, with the extent and pace of change needed soon there will be limited room for the traditional players. IOCs are evolving to become IECs, adopting a holistic business model to energy production and consumption, and recognizing the importance of diversification into low carbon and green energy in the face of the energy transition. Service providers are also evolving to meet the challenge, embracing the change and supporting the IECs along the way.

The adoption of digital technologies and step-change approach to innovation is a major driver to support the transformation. Energy companies recognize the power of digital technologies and have embraced the digital revolution to optimize their operations and services, enhance efficiency, and reduce costs. IoT, however not a particularly new concept, is being adopted more rapidly in the wake of the ET. IoT allows energy companies to collect data from their equipment and operations in real time, enabling them to monitor and optimize their operations. Big data analytics and AI are crucial technologies that work hand in hand with IoT. Armed with these digital tools, Energy companies can unlock new avenues for efficiency and operational effectiveness and drive new growth. Vast amounts of data can be collected from operations, customers, and other sources. Advanced analytics can extract insights that help make better decisions. By leveraging big data analytics and AI, energy companies can improve forecasting accuracy, optimize supply chains, and identify new business opportunities among many other benefits.

The rapidly changing energy landscape is no longer exclusive to traditional players and has paved the way for new players to enter the market. Titans of the technology world, like Google and Apple, have entered the energy sector, bringing with them disruptive technologies and novel ideas. To remain relevant and competitive energy companies must be agile and nimble, leveraging their vast experience and resources and embrace change to stay in the game. A marriage of traditional energy expertise and cutting-edge technology is crucial for navigating the complex terrain of the energy transition. The ET has opened up new business opportunities that go beyond traditional oil and gas exploration and production. Energy companies are diversifying their businesses and investing in innovation and digital solutions, and new opportunities, such as energy storage, EV charging infrastructure, and smart grid technologies. This enables them to future proof their portfolios by entering new up and coming markets and potentially be more competitive.

The changes also bring new ways of working and new ways to buy energy projects and services. Traditional contracting approaches in the oil and gas sector historically tended to be driven by lowest price wins, with a heavy weightage of the risk piled onto the contractor. The ET era demands a more robust and efficient way to deliver projects with an appreciation of the uncertain path ahead, which has resulted in a shift toward more progressive contracting methods, such as collaborative contracting. In addition, energy companies are partnering with governments, academic institutions, and other stakeholders to balance the risk, promote innovation, and develop more effective solutions to global energy challenges. Collaboration is crucial in the journey toward a sustainable energy future, and energy companies are leading the charge in this regard.

With the multifaceted changing landscape, it is necessary to recognize the importance of nurturing a skilled and agile workforce that are equipped deal with the ET change journey. This change demands a fresh mindset and operating philosophy comfortable to deviate from the norm, leaving the traditional approach behind. Investment in the workforce is of paramount importance, with upskilling centered around targeted training and development programs to foster a new generation of professionals with the expertise, digital acumen, and adaptability needed to thrive in the face of disruption. This in turn requires a new breed of leaders who embrace change and innovation, and recognize the value of shared objectives and partnerships to delivery successful projects and services.

References

[1] TotalEnergies, TotalEnergies Refinery of the Future, 2023.
[2] BP Press Release, BP and Beyond Limits Form Strategic Partnership to Boost Performance, 2023.
[3] Microsoft News, Shell and Microsoft Form Alliance to Help Address Carbon Emissions, 2019.
[4] TotalEnergies Website Article, TotalEnergies Ventures Commits 400 Million in Venture Capital to the Development of Start-Ups in the Renewable and Energy Sectors, 2019.

CHAPTER THREE

Collaborative contracting

Beyond traditional limits

Renowned as a powerhouse in the global energy landscape, the Gulf Cooperation Council (GCC) comprising countries Saudi Arabia, the United Arab Emirates, Oman, Kuwait, Bahrain, and Qatar has gained prominence through its trailblazing megaprojects that defy the limits of achievement in the oil and gas sector. Exemplary endeavors like Saudi Arabia's Khurais Oil Field Expansion, which aims to elevate production capacity by 1.2 million barrels per day. NEOM project in Saudi Arabia, which is a futuristic, sustainable city being built from scratch in the northwestern part of the country, near the borders with Jordan and Egypt, with an estimated cost around $500 billion, making it one of the largest and most ambitious mega projects in the world demonstrate the GCC's unwavering commitment to innovation and maximizing energy resources. Remarkably, as of 2020, the region boasted a staggering $2.9 trillion worth of major projects in the pipeline or already underway, unequivocally underlining its immense potential for future groundbreaking initiatives.

Fueled by an unyielding pursuit of energy excellence, the GCC has cemented its position as a global industry leader. Historically, a dominant force in the fossil fuel market owing to its substantial oil and gas reserves, and the GCC nations now embrace the imperative of diversification and sustainability. Actively charting a course toward cleaner and renewable energy sources, the region has set ambitious targets such as Saudi Arabia's vision to achieve 50% clean energy by 2030 and the United Arab Emirates' pioneering investments in solar and wind power. Leveraging their considerable financial resources, technological acumen, and strategic collaborations, the GCC countries strategically position themselves at the vanguard of the global energy transition. Their resolute dedication to propelling sustainable innovation and meeting future energy demands ensures their pivotal role in shaping the trajectory of the entire industry.

Powering Through the Transition
https://doi.org/10.1016/B978-0-323-91754-4.00003-0

However, GCC's pursuit of its developmental project agenda has exposed the urgent need for a paradigm shift in project delivery methodologies. Traditional contracting and project execution models have consistently demonstrated major drawbacks, such as creating bottlenecks that lead to considerable waste, fostering adversarial behaviors that promote disputes, and compromising on quality in favor of lowest cost, all of which contribute to delays and cost overruns. Given the increasing financial pressures, project sponsors and clients now demand greater social and economic returns on their investments to address the requirements of rapidly expanding populations and achieve national goals related to diversification and competitiveness.

One of the main factors contributing to project inefficiency on a global scale is the absence of collaboration and shared objectives among stakeholders. Problems, such as vague project specifications, eleventh-hour design modifications, inadequate or poorly documented communication, and reliance on outdated information, can quickly snowball into delays, increased costs, and contractual disagreements. Consequently, costly and contentious legal proceedings arise, fostering mistrust and eroding team cohesion within projects. Resources become entangled and settling late payments can take years.

A 2020 industry survey conducted by Middle East Economic Digest (MEED) discovered that waste levels in oil and gas projects within the GCC region could reach as high as 25%–30%. Other research suggests that time and cost overruns are rampant throughout the region's project industry, a trend reflected worldwide. The growing scale and scope of developments also increase project complexity, leading to heightened commercial and technical risks and an expanding roster of project participants [1].

At this juncture, it is important to elaborate on the term "traditional contracting" since it will be referred to in the following sections. It is a well-known expression and a widely used project delivery method in the construction and other industries. It is often referred to as design-bid-build or sequential contracting and is recognized for its linear approach, separating design and construction phases, and the competitive bidding process. While traditional contracting has been the standard approach for many years, it is increasingly being seen as outdated in certain contexts, particularly for complex or fast-track projects that require a higher degree of collaboration and integration between design and construction teams. In recent years, there has been a growing recognition of the limitations and challenges associated with traditional contracting, such as a focus on the lowest cost, misaligned priorities, fostering adversarial relationships, and limited collaboration. This has led to the development and adoption of alternative project delivery methods like design-build, Integrated Project Delivery (IPD), and collaborative contracting models, which are predicated on collaboration, shared risks and rewards, and a more integrated approach to project management [2].

Some of the challenges associated with traditional contracting include the following:

In traditional contracting, the common practice involves a competitive bidding process, where contractors submit their bids based on the design documents. The owner then selects the contractor with the lowest or most suitable bid. However, this approach can sometimes lead to issues. Contractors may feel compelled to underbid in order to secure the contract, which can potentially compromise the quality, safety, or timely completion of the project. Moreover, cash flow can become a challenge for contractors working under traditional contracts, as payment terms may be tied to project milestones or require substantial completion before payment is made. This financial strain can impact the contractor's ability to deliver the project on schedule and within the allocated budget. Additionally, traditional contracting often involves separate contracts between the design and construction parties, resulting in limited collaboration. This lack of collaboration can lead to an increased number of change orders, claims, and disputes, ultimately driving up project costs, causing delays, and fostering a negative working environment.

The linear process and distinct roles often foster an environment where trust is lacking, communication is poor, and individual interests take precedence over the overall project success. The limited collaboration between design and construction parties further exacerbates these issues, as they work separately without significant interaction throughout the project lifecycle. This lack of synergy hampers innovation and stifles opportunities for value engineering or process improvements, potentially leading to suboptimal outcomes.

It is important to note, however, that traditional contracting is not necessarily outdated or unsuitable for all types of projects. In some cases, traditional contracting can still be an effective and appropriate project delivery method, especially less complex projects, where the design and construction requirements are relatively straightforward, where the scope and interfaces are clearly defined. However, these challenges have led many organizations to explore alternative project delivery methods, such as design-build, IPD, or collaborative contracting models with shared risks and rewards, and a more integrated approach to project management. These alternative methods aim to address the limitations of traditional contracting and create a more efficient, positive, and successful project environment.

In 2018, Deloitte published a report on oil and gas contracting, which highlighted that the adversarial nature of traditional contracting in the industry has been a significant factor contributing to inefficiencies, project delays, and cost overruns. The report shed light on various aspects of oil and gas contracting that have led to project delays, disputes, and litigation, ultimately affecting the overall performance of the industry. The report went on to emphasize the adversarial nature of "traditional contracting" in the oil and gas industry, which often stems from a lack of trust and collaboration between various stakeholders. Traditional contracting models, such as lump sum contracts, tend to prioritize risk allocation and cost minimization over collaboration and innovation. This can inadvertently create an environment where parties are more concerned with protecting their own interests than working together to achieve the project's objectives. Several factors contribute to the adversarial nature of traditional contracting [3]:

- **Poorly defined scopes of work**
  Ambiguous or incomplete project specifications can lead to misunderstandings between parties, resulting in cost overruns and project delays. This lack of clarity can cause disputes and create an environment where stakeholders are more likely to engage in litigation to resolve conflicts.
- **Inflexible contract terms**
  Rigid contract terms that do not allow for adjustments as the project progresses can lead to conflicts between parties. As the complexity of oil and gas projects increases, the need for flexibility in contract terms becomes more critical to accommodate unforeseen changes and maintain project efficiency.
- **Inadequate risk allocation**
  An imbalanced allocation of risks between parties can create tension and mistrust, ultimately leading to disputes and litigation. For example, if a contractor is unfairly burdened with excessive risk a claims-culture tends to develop where contractors may resort to legal action to protect their interests.
- **Insufficient communication**
  A lack of open and transparent communication between parties can hinder collaboration and lead to misunderstandings, disputes, and project delays. Establishing clear channels for communication and fostering a culture of transparency can help to mitigate these issues.

- **Lowest cost**

  Typically lump sum turnkey contracts tend to drive a "lowest cost" approach due to the characteristic competitive bidding process, which also tends to result with the contractors assuming the majority of the project risk. Clearly this approach fails to promote effective collaboration between parties involved. Lowest bidders and the adversarial nature of lump sum contracts makes it increasingly difficult for project parties to mitigate design and construction risks. As a result, clients often face late deliveries, budget overruns, and quality issues, while contractors and consultants grapple with the commercial risks that can transform projects into loss-making ventures. This unsustainable model often leads to companies prioritizing cost reduction over delivering high-quality projects for clients and end users, further perpetuating inefficiencies within the industry.

One of the key concerns for companies working in the in a traditional contracting environment is payment delays, which require contractors and suppliers to finance large portions of the work as they navigate a complex chain of approvals and certifications before receiving payment. The impact on business cash flow discourages companies from making additional efforts that go beyond their contractual commitments, even when such efforts could benefit the overall project. The current contracting model is no longer tenable, and a shift toward a more collaborative and efficient approach is necessary for the long-term success.

To address these concerns, more pragmatic and collaborative approaches are needed. Energy companies operating traditional contracting approaches must be willing negotiate and balance risk appropriately to motivate service providers and contractors and to align on common priorities, primarily the safe and successful execution of the project. A long-term investment in relationships between service providers and clients is needed rather than just focusing on short-term gains. There are a wide range of contracting approaches that are available, and collaborative contracting approaches are becoming more accepted, especially in the GCC, such as IPD or Engineering, Procurement, and Construction Management (EPCm) contracts, where the risk profile is more evenly balanced between contractor and client.

The recognition among clients that traditional contracting approaches have major shortcomings has become more prevalent in recent years. These methods often disproportionately apportion risks to contractors, which likely leads to adversarial relationships between contractors and clients that hinder project success. Such an approach aligns with an outdated management and leadership style that emphasizes control and rigidity rather than collaboration and adaptability. As a result, there is a pressing need to move beyond these antiquated practices in order to foster more productive and harmonious partnerships within the energy sector.

The shift toward more progressive contracting methods, such as collaborative contracting, requires a new breed of leaders who embrace change and innovation. These leaders recognize the value of shared objectives and mutual trust, and they understand that a project's success is directly tied to the strength of the relationships forged among all stakeholders. By adopting a more inclusive and flexible leadership style, these modern leaders foster an environment, in which all parties are encouraged to contribute their expertise and work together toward common goals. This approach not only mitigates adversarial dynamics but also cultivates a sense of shared responsibility and accountability, ultimately resulting in more effective and efficient project outcomes.

As the energy landscape continues to evolve, the emergence of forward-thinking leaders is essential in driving the adoption of progressive contracting methods. These new leaders must possess strong communication skills, a willingness to learn, and the ability to empower their teams to navigate the complexities of modern projects. Actively fostering collaboration, embracing change, and promoting a culture of continuous improvement enables these leaders to set the stage for a future in which the ET thrives, and all stakeholders benefit from more equitable and successful partnerships.

Contracting approaches

The foundation of a relationship between a contractor and a client is the contract itself, making it important to understand the core principles of a contract as part of this book. A contract can be defined as a legally binding agreement between two or more parties, allocating responsibility and risk. Standard contracts offer several advantages, such as well-established clauses that simplify interpretation and minimize the need for expensive litigation.

One of the most prevalent international standard contracts is issued by the Federation Internationale des Ingenieurs Conseil (FIDIC). These contracts recognize that circumstances differ across the world and allow for modifications to accommodate them. The first part, the general conditions, is intended to be universal and should not be altered. The second part, the conditions of application, is specifically tailored for each contract. Sample clauses are provided to help users customize these conditions to suit each unique situation.

For large-scale projects with significant responsibilities and risks, detailed FIDIC contracts or similar standard contracts are both necessary and justified. However, for smaller scopes of work carried out by labor-based contractors, it is debatable whether intricate contracts are truly required to protect the interests of a relatively affluent client against a less privileged small entrepreneur. All contracts inherently involve risk. One of the fundamental reasons clients employ contractors is to transfer this risk, and the contractor's reward for bearing this risk is the profit expected beyond estimated costs and a reasonable commercial return. The greater the risk, the higher the required and anticipated profit. If the expected risk-to-reward ratio is too high, a competent contractor will avoid the project entirely.

Before deciding on the optimal level of risk transfer, it is crucial to define risk and examine methods for analysis and assessment. Remember that risk can rarely be eliminated entirely from any human endeavor, as it is inherent in committing current resources to achieve future benefits. Therefore, no contractual system, no matter how complex, can guarantee a risk-free project. However, risk can be identified, analyzed, and mitigated.

Clients should especially note that minimizing project risk benefits all parties involved. Once this is achieved, the next step should be determining the amount of residual risk to transfer to the contractor. Transferring risk comes with a premium, so it is generally best to transfer the lowest realistic level of risk while ensuring the contractor remains accountable for performance. This approach balances the need to protect the client's interests while maintaining a fair and efficient contractual environment for all stakeholders.

There are a multitude of different contract models that can cater to the diverse client preferences and requirements of project delivery. These models serve to delineate the relationship between clients and contractors, with each model offering distinct features, benefits, and drawbacks based on project specifications, risk distribution, and the desired degree of control.

Given the unique nature of each project, no one-size-fits-all approach exists for choosing the most suitable contract model. Making an informed selection among the available options requires a comprehensive understanding of various aspects, such as timing and budget expectations, the extent of project definition, ownership of technological know-how, desired control over the project, the proficiency and availability of an in-house project team, and the capacity to manage guarantee obligations, among other elements.

To ensure the successful evaluation and selection of a contract model that aligns with an organization's objectives, a trusted partner is often required. Organizations must accurately assess their needs and establish a productive working relationship with contractors. This approach will enable them to navigate the complex landscape of contract models and optimize the outcome of their projects. Several prevalent contract models are presented next.

Lump sum or fixed-price contract

Traditional contracting, often referred to as a lump sum or fixed-price contract, is an arrangement where the contractor commits to completing a project at a predetermined cost. The lump sum contract provides clients with cost certainty as the contractor assumes the responsibility for any cost overruns. This contract model is suitable for projects with well-defined scopes, precise specifications, and predetermined schedules, although it carries varying levels of risk for both the contractor and the client.

For clients, the lump sum contract allows effective budget management and helps avoid unexpected expenses. With the contractor obligated to adhere to the fixed price, clients can focus on other project aspects without worrying about escalating costs. Conversely, contractors have the opportunity to increase their profit margins by delivering the project under budget and ahead of schedule. However, accurately estimating project costs and accounting for potential changes or contingencies can pose challenges for contractors.

It is important to note that while the lump sum contract offers several benefits, it may not be suitable for every project. Projects with undefined or frequently changing scopes can lead to disputes, delays, and cost overruns under this contract model. In such cases, alternative contract models that provide more flexibility, such as cost reimbursable or time and material contracts, may be more appropriate. Choosing the right contract model requires careful consideration of the project's characteristics and objectives to ensure optimal outcomes.

Unit rate contract

The concept of a unit rate contract revolves around predetermined rates for different work components, where the contractor's payment is determined based on the actual quantity of work completed. This contract model is particularly suitable for projects where the work quantities are uncertain or the scope cannot be precisely defined. In terms of risk, unit rate contracts carry a moderate level of risk for both the contractor and the client, as they share the risk of potential cost overruns.

Unit rate contracts offer flexibility in accommodating changes to the project's scope without requiring extensive renegotiations. This adaptability is particularly valuable when the project's nature or requirements are subject to variation, or when the site conditions are not fully known in advance. By aligning the contractor's payment with the actual work accomplished, unit rate contracts foster a collaborative environment and incentivize the contractor to deliver high-quality work within the specified timeframe.

However, it is crucial for both parties to closely monitor the project's progress and costs, as the shared risk in unit rate contracts can potentially lead to disputes concerning work quantities or rates. To mitigate such conflicts, the contract should clearly define the methods for measuring completed work and specify the procedures for adjusting rates when necessary.

Ultimately, the unit rate contract model promotes a balanced distribution of risk and encourages collaboration between the contractor and the client. To ensure the success of projects executed under unit rate contracts, both parties must approach the project with transparency and a shared commitment to its successful completion.

Cost reimbursable contract

The cost reimbursable contract is an agreement in which the client consents to compensate the contractor for the actual expenses incurred throughout the project, typically accompanied by a predetermined fee or percentage markup to cover overhead costs and profit. This contract model affords a level of flexibility in the project's scope, granting the client greater control over project-related decisions. In terms of risk distribution, the contractor experiences minimal risk, while the client assumes a higher degree of risk.

The cost reimbursable contract's inherent flexibility is one of its most significant advantages, particularly for projects with uncertain or evolving scopes. As the project progresses, the client can make adjustments without the need to renegotiate the entire contract. This adaptability can lead to a more collaborative relationship between the client and the contractor, fostering innovation and enabling the project to achieve better results. However, the primary drawback of this model is the increased financial risk for the client, as they are responsible for covering any cost overruns. Consequently, clients must actively monitor project expenses and maintain open lines of communication with the contractor to ensure effective cost management.

Despite the potential drawbacks, a cost reimbursable contract can be an ideal choice for projects that demand a high degree of adaptability, such as research and development initiatives or complex construction projects. To mitigate the associated risks, clients should implement robust project management practices, including ongoing cost tracking, regular progress reviews, and clear communication with the contractor. By embracing these strategies, clients can leverage the benefits of a cost reimbursable contract while minimizing financial risks, ensuring that the project achieves its objectives within the desired time frame and budget constraints.

Engineering, Procurement, and Construction (EPC) contract

An EPC contract represents a comprehensive project model where the contractor assumes responsibility for all aspects, including design, EPC. With an EPC contract, the client benefits from having a single point of contact, streamlining communication, and reducing the number of interfaces. This model is particularly suitable for large and complex projects that have well-defined outcomes. In terms of risk, the contractor bears a high level of risk for the project's success, while the client assumes a relatively low level of risk (Fig. 3.1).

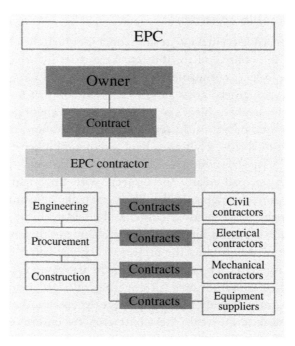

Fig. 3.1 IECs.

An EPC contract can provide clients with the advantage of consolidated liability in the event of warranty cases. By entrusting the entire project to the contractor, clients can limit their direct involvement, allowing them to focus on other aspects of their business. However, it is important to consider that the process of obtaining an offer for an EPC contract can be time-consuming, typically taking 2–4 months as the contractor monitors the market to fix the price. Additionally, the project cost for an EPC contract may be 20%–30% higher compared to an EPCm project approach. This is due to the numerous assumptions and provisions made in the contract to cover potential risks. Once the contract is signed, the contractor assumes a monopoly position, and the project execution becomes entirely their responsibility. Consequently, the client's capital expenditure is fixed from the start. While this limits the client's involvement in the process, any changes requested by the client during the project may lead to additional expenses.

Therefore, it is crucial to define technical requirements as thoroughly as possible during the initial stage and have them clearly stipulated in the EPC contract, including guarantee parameters, test procedures, and required standards. Failure to do so may result in exceeding the budget and project deadlines.

EPC contracts can be structured as either reimbursable or lump sum contracts, depending on the agreement between the client and the contractor. In a reimbursable EPC contract, the contractor is reimbursed for the actual costs incurred during the project, including labor, materials, equipment, and other expenses. The client pays the contractor based on the receipts and invoices provided, and the contractor typically receives a fee or markup on the reimbursed costs. Reimbursable contracts provide more transparency in terms of cost breakdown and can be beneficial when the project scope is uncertain or subject to changes. On the other hand, a lump sum EPC contract involves a fixed price agreed upon between the client and the contractor for the completion of the project. Under this arrangement, the contractor assumes the responsibility for managing costs and any potential cost overruns. Lump sum contracts provide cost certainty for the client, as they know the total price upfront, and the contractor takes on the risk of completing the project within the agreed budget.

The choice between reimbursable and lump sum EPC contracts depends on various factors, including the project's complexity, the level of project scope definition, and the risk appetite of both the client and the contractor. Some projects may start as reimbursable contracts during the initial phases when the scope is not fully defined and then transition into a lump sum contract once the scope becomes clearer.

EPCm contract

An EPCm contract is a modified version of the traditional EPC model. In an EPCm contract, the contractor is responsible for managing the project's EPC, but the client retains some level of control over the design and procurement processes. This approach provides greater flexibility for the client and shares the risk between both parties involved. The risk level is considered moderate for both the contractor and the client (Fig. 3.2).

One of the key advantages of an EPCm contract is its flexibility in terms of project execution timing and budget. Compared to an EPC contract, the client can expect to receive a proposal from the service provider in a shorter timeframe, typically 2–4 months faster. Additionally, the overall project costs are likely to be lower as there is no need for the supplier to include additional funds for risk mitigation. Furthermore, before signing the contract(s) for construction works, the client faces minimal financial risks in case of project cancellation. The step-by-step nature of the project allows the client to introduce design changes or refine the strategy during project execution, providing greater adaptability. Moreover, this approach enables the client to engage professional companies specializing in different fields, including design and construction.

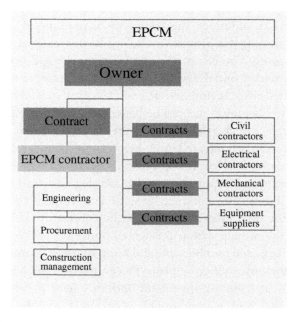

Fig. 3.2 IECs.

However, it is important to note that with an EPCm contract, the responsibility is shared between the designer and the contractor. This coordination between multiple parties requires a competent team from the client's side to effectively oversee the contractors and ensure timely and accurate technical decision-making. Effective communication and collaboration between the client, contractor, and designer are crucial to successfully implement an EPCm contract. By having a qualified team in place, the client can effectively manage and control the project, ensuring that all parties are aligned and working toward the project's objectives.

Project Management Contract (PMC)

In a PMC contract, the contractor assumes the role of the client's agent or representative, taking charge of overseeing and managing the various elements of the project, including design, procurement, and construction. While the contractor does not directly perform the construction work, they are responsible for coordinating and managing the activities of subcontractors. This contract model allows the client to maintain control and decision-making authority over the project while benefiting from the contractor's expertise. The risk level is generally low for the contractor, as they are primarily responsible for project management, and high for the client, who retains the ultimate responsibility for the project's success.

The PMC approach is particularly beneficial in cases where the engineering requirements demand-specific process capabilities that are not easily found in the market. By splitting the responsibility for engineering and project management functions between two contractors, the PMC partner can act as the client's representative during both the design and construction phases. During the design phase, they perform design coordination and evaluation, ensuring the engineering aspects are aligned with the project objectives. In the construction phase, the PMC partner takes on the role of construction management on-site, coordinating the activities of subcontractors across various disciplines. This model is well-suited for projects with a multicontracting approach, as the PMC partner assumes the role of the general contractor, managing and coordinating the activities of subcontractors in civil, mechanical, electrical, and other disciplines. With the PMC contract, the client retains the freedom to implement changes at any stage of the project, with the PMC service provider managing and accommodating those changes. However, it is important to note that any changes made may have a significant impact on the project's time and budget.

For clients who prefer fixed costs but still require professional external management, combining an EPC contractor with a PMC company can be a viable option. This approach allows the client to benefit from the cost certainty provided by the EPC model while leveraging the project management expertise of the PMC contractor. By merging these two approaches, clients can have the assurance of fixed costs and professional management throughout the project's lifecycle. This combination offers a flexible solution that caters to the specific needs and preferences of the client (Fig. 3.3).

In today's evolving project landscape, the selection of an appropriate contract model has become increasingly critical to determine project success. Clients and contractors recognize that one size does not fit all when it comes to contracts. The wide range of contract models available, such as EPC, EPCm, and PMC, allows for a more tailored approach to suit specific project types and objectives. Each model carries different risk levels and offers unique benefits, such as cost certainty, flexibility, or increased control.

As the industry embraces collaborative contracting, stakeholders are realizing the benefits of early involvement, integrated teams, and shared responsibilities. These models prioritize collaboration and alignment of objectives from the outset, allowing for improved project planning, risk management,

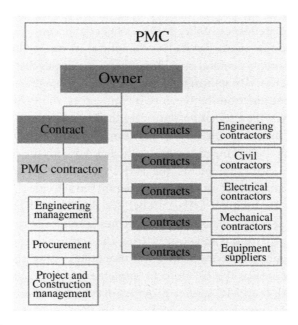

Fig. 3.3 IECs.

and problem-solving. By fostering a spirit of partnership, clients and contractors can better navigate challenges, adapt to changing project requirements, and achieve better overall outcomes.

Changing the conversation—Collaborative contracting

The global energy sector stands at a critical juncture, facing an array of unprecedented challenges. Uncertain market conditions, escalating regulatory demands, and the ever-present need for operational efficiency have prompted a growing realization that the traditional contracting approach is no longer viable. Consequently, the industry is undergoing a transformative shift toward collaborative contracting, recognizing it as a catalyst for innovation, streamlined project delivery, and enhanced value for all stakeholders involved.

The traditional contracting model, characterized by adversarial relationships and a focus on securing the lowest bid, has been a significant contributor to inefficiencies, delays, disputes, and litigation of historical project delivery. Forward-thinking energy companies are acknowledging the imperative to depart from this outdated approach and embrace a more collaborative mindset. Fostering a culture of trust, shared goals, and open communication promotes collaborative contracting to pave the way for enhanced project outcomes, increased productivity, and reduced waste.

The transition to collaborative contracting is crucial for the long-term success of the energy sector as well as essential for attracting investment, promoting economic growth, and ensuring environmental sustainability. Adopting a more cooperative approach allows companies to better manage risks, leverage the collective expertise of their project partners, and foster a culture of continuous improvement. The industry's future depends on its ability to embrace change and adapt to the evolving demands of the market, making the shift toward collaborative contracting a crucial step on this journey.

Project teams structured to collaborate, rather than operate in silos, are more likely to deliver superior outcomes, improve productivity, significantly reduce abortive work and waste, and foster a culture that encourages joint problem-solving and innovation in project delivery. In this approach, project partners are selected based on the value they contribute to a project, considering factors, such as cultural approach, corporate governance, safety records, commitment to continuous improvement, and innovation. The success of each project team member is directly linked to the overall project's success, with predefined common targets measured using key performance indicators. While collaborative contracts can be applied to various project types, they yield the greatest benefits when used for complex, medium-to-long-term projects involving multiple stakeholders and subject to various key risks that must be identified and effectively managed from the start.

Elements of collaborative contracting

Collaborative contracting is a relationship-based approach to procurement and project management that focuses on cooperation, trust, and shared objectives between all parties involved. This approach aims to create a more integrated and mutually beneficial working environment, leading to better project outcomes and reduced risk. Some examples of collaborative contracting models are as follows:

- **IPD:** IPD is a project delivery approach that fully integrates project teams, including the owner, engineers, contractors, and subcontractors. It emphasizes collaboration from the early stages of the project and often involves shared risk and reward structures to align all parties' interests.
- **Alliance contracting:** In this model, the owner, contractors, and other key stakeholders form an alliance to deliver the project. The alliance typically shares risks and rewards based on the project's outcomes, fostering trust and cooperation among parties. This approach often involves setting common goals and promoting transparency in decision-making and information sharing.
- **Partnering:** Partnering is a more informal approach to collaboration, focusing on building trust and cooperation between project participants. It often involves signing a partnering charter or agreement that outlines the project's goals and the principles of cooperation, communication, and dispute resolution. While it may not involve the same level of risk sharing as other models, partnering can still foster a collaborative environment.
- **Joint ventures:** In a joint venture, two or more parties come together to form a new legal entity, sharing resources, risks, and rewards for the project. Joint ventures can be effective in collaborative contracting, as they encourage cooperation and shared goals among the participating organizations.
- **Framework agreements:** These are long-term agreements between clients and suppliers that provide the framework for future projects. They typically outline the terms, conditions, and pricing for a series of projects

or services, which can be called off as required. Framework agreements can promote collaboration by providing a stable and predictable working relationship between parties.

- **Target cost contracts:** This type of contract sets a target cost for a project, and parties share any cost overruns or savings based on an agreed-upon formula. This approach encourages collaboration by aligning the interests of both the client and contractor to complete the project within the target cost.
- **Early contractor involvement:** In this model, the contractor is engaged during the design phase of the project. This allows for more effective collaboration and coordination between the design and construction teams, leading to improved project outcomes and reduced risk.

Each of these collaborative contracting models emphasizes shared goals, open communication, and cooperation among all parties involved in a project, ultimately leading to better outcomes and a more positive working environment.

An example to illustrate the advantages of an alliance contract when deployed correctly is the delivery of the Reliance Industries Limited, Jamnagar Refinery in India. The project involved the construction of one of the world's largest and most complex refineries, capable of processing various types of crude oil and producing a range of petroleum products.

The project was executed through an alliance contracting model, which brought together the project owner, Reliance Industries Limited, and several international engineering and construction firms, such as Bechtel, UOP, and Foster Wheeler, among others. The alliance was formed to manage and deliver the design, procurement, and construction of the refinery, aiming to achieve tight deadlines and cost objectives while maintaining high safety and quality standards.

The alliance partners worked together in a collaborative environment, sharing risks and rewards, and leveraging the expertise of each partner. The project was delivered ahead of schedule and within budget, and the refinery began producing high-quality products that exceeded industry standards. The successful completion of the Jamnagar Refinery can largely be attributed to the effective collaboration and shared commitment of the alliance partners to achieve common goals, which demonstrates the potential benefits of alliance contracting in delivering large, complex infrastructure projects by fostering a collaborative environment, aligning the interests of all parties, and leveraging the combined expertise of the project team.

Consequences of doing nothing

There are several reasons for this shift in client views on contracting models. One major factor is the frustration of the currently levels of successful projects delivered under traditional contract approaches; on top of this, energy projects are becoming more complicated, especially when the unknowns of energy transition era are considered. This has led clients to seek more control over the design and procurement process. This allows clients to manage the project risks more effectively and to ensure that the project aligns with their specific needs and goals. Additionally, as the construction industry has become more competitive, clients have sought to reduce the risks associated with cost overruns and delays, which are more likely to occur under the traditional EPC lump sum model.

Another reason for the shift toward risk-shared models is the increasing focus on sustainability and environmental responsibility. Under the traditional EPC lump sum model, the contractor is often motivated to cut costs, which can lead to the use of cheaper, less sustainable materials and methods. With the EPCm and PMC models, the client has more control over the design and procurement process, which allows them to ensure that the project is built using sustainable materials and methods.

According to a report by Accenture, "new contractual models are required that recognize the changing market conditions, support collaboration, and promote innovation." This means that contracts must be structured in a way that promotes collaboration and innovation, rather than just maximizing profits for one party. In addition, a more collaborative approach to contracting requires a high level of trust, communication, and transparency. This requires a cultural shift within the industry, where contractors and clients work together as partners to co-create solutions that benefit everyone involved [4].

The Papua New Guinea (PNG) Liquefied Natural Gas (LNG) project is a prime example of a successful collaborative approach to contracting. The project, led by ExxonMobil in partnership with JGC Corporation and other stakeholders, involved the construction of a world-class LNG processing facility and infrastructure to produce and export natural gas from the Hela, Southern Highlands, and Western provinces of PNG. The project required a highly collaborative approach, as it involved complex engineering, environmental, and logistical challenges, including constructing pipelines through remote and rugged terrain, accommodating local communities' needs, and adhering to stringent safety and environmental standards. To address these challenges, ExxonMobil and JGC Corporation adopted a more collaborative approach to contracting.

The two companies worked together to negotiate a balanced contract that emphasized risk sharing and collaboration on technical solutions. ExxonMobil and JGC Corporation maintained open lines of communication throughout the project, fostering a culture of trust and transparency. This allowed the companies to address challenges proactively and find mutually beneficial solutions. The contract included provisions for sharing risks and rewards, aligning the interests of both companies and encouraging them to work together to achieve the project's goals. They collaborated on finding innovative technical solutions to the project's unique engineering and logistical challenges. This joint problem-solving approach enabled the project team to overcome obstacles and improve overall project performance. As a result of this collaborative approach to contracting, the PNG LNG project was a success. In 2014, the project began exporting LNG, making PNG the world's newest LNG exporter. The success of the PNG LNG project demonstrates the potential benefits of adopting a more collaborative approach to contracting, particularly for complex, large-scale projects with multiple stakeholders and unique challenges [5].

"One team"

In 2013, BP entered into an agreement with its main drilling contractor, Seadrill, for the operation of the West Vela rig in the Gulf of Mexico. This traditional contract had several key challenges, including misaligned incentives and a lack of collaboration. BP sought to maximize efficiency and reduce costs, while Seadrill aimed to maximize revenue, potentially leading to higher costs for BP. This misalignment, coupled with the rigid nature of the contract, which contained specific requirements and metrics, limited the companies' ability to adapt to changing circumstances or pursue innovative solutions. Furthermore, the contract's rigidity and lack of collaboration could have resulted in communication breakdowns, inefficiencies, and missed opportunities for continuous improvement.

Another significant challenge was the unbalanced approach to risk allocation. Both parties attempted to shift potential liabilities to the other, creating a tense atmosphere that reduced trust and hampered the development of a genuine partnership. In addition to this, the traditional contract relied on various performance metrics to assess the success of the relationship.

However, these metrics may not have accurately captured the full scope of performance, leading to an incomplete understanding of the partnership's effectiveness.

To overcome these challenges, BP and Seadrill had a reset moment, adopted an entirely new approach and coined the term the "One Team" approach, which emphasized aligning incentives, fostering collaboration, and creating a more flexible and adaptive working relationship. This new approach helped address many of the issues faced under the traditional contract, ultimately leading to improved performance and a stronger partnership between the two companies. The safety performance improved, increased operational uptime, both companies experienced cost savings and the rig's operational efficiency reached over 95%. By focusing on shared goals and establishing governance structures that kept both parties' expectations and interests aligned, the "One Team" approach proved to be a more effective way of managing complex relationships. This example demonstrates how shifting from traditional transactional contracts to performance-based, collaborative arrangements can enhance relationships and outcomes in the oil and gas sector.

This example is not unique. Companies in the oil and gas sector, as in other industries, recognize the importance of suppliers in reducing costs, increasing quality, and driving innovation. They talk about the need for strategic relationships with shared goals and risks, but during contract negotiations, they often default to adversarial mindsets and transactional contracting approaches. They focus on every possible scenario and try to put everything in black and white. Various contractual clauses are used to gain the upper hand, but these tactics often lead to false security and negative behaviors that undermine the relationship and contract.

The remedy is adopting a different kind of arrangement and leadership, such as formal relation-based contracts that specify mutual goals and establish governance structures to keep the parties' expectations and interests aligned over the long term. These contracts are designed to foster trust and collaboration, particularly in complex relationships where it is impossible to predict every possible scenario. More organizations, including oil and gas companies, are successfully using this approach to achieve better outcomes in their supplier relationships.

Shading

In 2008, Nobel Prize winner Oliver Hart for his work in Contracts, delved into the complexities of incomplete contracts and the impact of perceived fairness on parties' behavior, an idea termed "shading." He introduced a model that highlights the psychological aspects of contractual relationships, particularly the importance of reference points in determining fairness perceptions. When the terms of a contract deviate from these reference points, it may lead to feelings of unfairness and consequently, shading behavior, where parties intentionally underperform or reduce their efforts.

Hart emphasized the challenges posed by incomplete contracts, which are unable to address every possible contingency. This inherent incompleteness leaves room for shading behavior when unforeseen situations emerge. Hart argued that parties in a contractual relationship have certain expectations or reference points based on past experiences or industry norms, and deviations from these reference points can give rise to dissatisfaction and shading behavior. Shading behavior complicates contract negotiations and renegotiations, as parties may engage in strategic actions to achieve desired outcomes. The potential for shading behavior can even lead to inefficient outcomes, as parties may choose to punish the other party by underperforming or withholding effort, even at their own expense. The study by Hart and Moore underscores the significance of understanding and addressing the psychological dimensions of contracts to mitigate shading behavior and improve contractual outcomes.

To illustrate this idea, which many readers may have experienced, we consider an engineering and construction company (Company A) and a client (Client B) entering into an engineering and build contract. The contract specifies a fixed payment for the completion of the project within a certain timeframe. However, the contract fails to address specific contingencies, such as extreme weather conditions or delays in obtaining permits, which could affect the project's timeline.

As the project progresses, unexpected weather causes delays in construction. Company A requests an extension and additional payment to cover the unforeseen costs, but Client B refuses, arguing that it is Company A's responsibility to manage such risks. This disagreement leads to a deviation from the initially agreed-upon reference points, causing both parties to perceive the contract as unfair.

In response to the perceived unfairness, Company A engages in shading behavior by intentionally reducing the quality of materials used or cutting corners on safety measures, effectively lowering the overall quality of the finished structure. Meanwhile, Client B becomes uncooperative, delaying permit approvals and avoiding communication with Company A. Both parties suffer as a result: Company A incurs reputational damage and increased costs, while Client B receives a lower-quality oil facility that may require costly repairs in the future. This example illustrates how incomplete contracts and the deviation from reference points can lead to shading behavior, resulting in inefficient and suboptimal outcomes for both parties involved.

There are a variety of methods aimed at fostering healthier and more sustainable partnerships in contracting, and one such approach is the vested methodology for creating formal relational contracts, as outlined by Frydlinger, Hart, and Vitasek in their HBR article, "A New Approach to Contracts." This method combines the best elements of traditional contracts with relationship-building elements, promoting a "what's in it for we" mentality that encourages mutually beneficial outcomes.

Although relational contracts that depend on parties acting in their mutual self-interest are not new, the vested methodology addresses some of the shortcomings of informal "handshake" deals and other relational contract models. The formal relational contract, as proposed by the vested methodology, addresses these issues by incorporating both traditional contract components and relationship-building elements. These include a shared vision, guiding principles, and robust governance structures that help align the expectations and interests of the contracting parties. This approach focuses on developing strong relationships between the parties and that any conflicts that may arise are addressed through a collaborative problem-solving process. To date, the vested methodology has been employed numerous companies, with many reporting positive results, such as cost savings, improved profitability, higher levels of service, and better overall relationships [6].

Acknowledging that there is no universal solution when it comes to contracting, as techniques that are successful in one context might not be equally effective in others. As a result, it is important for project sponsors to evaluate the distinct characteristics of each project on an individual basis. This assessment will enable them to identify which components of a cooperative working approach can be effectively integrated and executed before settling on a particular strategy. To drive these progressive contract approaches forward, a dynamic and visionary leader capable of spearheading change is indispensable. We will explore the evolving role of the CEO in this new landscape in Chapter 10. Are traditional CEOs about to become a vestige of the past? It is becoming clear that the leadership style that served us in the preenergy transition era and digital era is no longer appropriate to serve us moving forward.

References

[1] J. England, 2018 Outlook on Oil and Gas, Deloitte, 2018.

[2] Accenture Article, Accenture, 'New Contractual Models Are Required to Navigate the Changing Market Conditions, Support Collaboration, and Promote Innovation', 2011.

[3] M. El Asmar, A.S. Hanna, W.Y. Loh, Quantifying performance for the integrated project delivery system as compared to established delivery systems, J. Constr. Eng. Manage. 139 (11) (2013).

[4] ExxonMobil's PNG LNG starts production ahead of schedule, Oil Gas J. (2014).

[5] D. Frydlinger, O. Hart, K. Vitasek, A New Approach to Contracts: How to Build Better Long-Term Strategic Partnerships, Harvard Business Review, 2019.

[6] Contracts as reference points, Theor. J. Econ. (2007).

CHAPTER FOUR

Embracing digital: A strategic imperative

Digitalization is not just a nice-to-have option for the energy sector. It is a must-have, a strategic priority, and a competitive differentiator.

CEO of Shell (2020)

The energy transition has ushered in a new era of opportunities and challenges for energy companies. To survive in this dynamic and uncertain landscape, it has become evident that energy companies must embrace the new age of digital and harnesses its transformative powers. The traditional approach that has served these companies well for decades is no longer sufficient to ensure survival and success.

Picture a future where power grids possess the intelligence to adapt dynamically to fluctuating energy demands, seamlessly integrating renewable sources to create a sustainable and reliable energy mix. Envision a scenario where advanced analytics, powered by the convergence of the Internet of Things (IoT), artificial intelligence (AI), and big data analytics, can predict and prevent disruptions before they even occur. Digitalization enables service companies to streamline and optimize their operations, such as enhanced customer experience, improve workforce efficiency through collaboration and knowledge sharing, and improve project delivery success through data collection and analytics to drive more robust and efficient decision-making, among many others. By leveraging IoT sensors and connectivity, energy companies can monitor equipment performance, predict maintenance needs, and prevent costly downtime. AI-powered analytics offer valuable insights into project risks and opportunities, allowing for proactive decision-making and mitigating potential delays or cost overruns. The integration of digital tools and platforms enhances collaboration among project stakeholders, promoting transparency, efficiency, and effective communication throughout the project lifecycle. This is the transformative promise of the new age of digital in the energy sector, which is already a reality.

However, for companies to harness the full potential of digitization and digital technologies, a fundamental shift in mindset and approach is required.

Powering Through the Transition
https://doi.org/10.1016/B978-0-323-91754-4.00001-7

Digitalization refers to enabling or improving operations processes by leveraging digital technologies and digitized data. Digital transformation is business transformation enabled by digitalization. It is not just a fancy buzzword; it has become a strategic imperative and a competitive differentiator. In this chapter, we will delve into the significance of digitalization and innovation in the energy transition and explore how companies must adapt to survive, and can even thrive in the new era by embracing the digital revolution and deploying fit-for-purpose digital solutions to tackle the challenges that lie ahead.

The new age of digital

We are now in the midst of a digital revolution and its impact can be felt far and wide. It has the potential to determine whether companies thrive or face the risk of not making it at all. While we have witnessed remarkable efficiency gains in the energy sector through the adoption of digital technologies since the downturn in 2014, we are still in the early stages of the digital revolution and the opportunities it presents. The widespread availability of increasingly affordable digital technology and low-code applications is already driving innovation throughout the oil and gas value chain. It has the potential to revolutionize field development, operational processes, and even the back-office functions that support core business operations. By leveraging digital technologies, companies can achieve radical efficiency gains and significant improvements in both their top and bottom lines. However, it is important to acknowledge that not all players will emerge victorious. The pace of the digital age surpasses that of all previous industrial revolutions, creating winners and losers in its wake.

Companies that demonstrate a willingness to innovate and invest in digital transformation stand to unlock substantial value. The digital cavalry is undoubtedly on its way, but it will not rescue everyone. Only those bold enough to embrace the digital revolution and adapt to its transformative power are likely to reap its rewards. This revolution is not a distant vision of the future; it is happening now, reshaping industries, and disrupting traditional business models.

So, what exactly is the "New Age of Digital"?

As we discussed in Chapter 2, to date there have been four industrial revolutions, which saw transformative periods in human history, ushering in significant changes to society and the economy. The First Industrial Revolution began in the late 18th century and was characterized by the

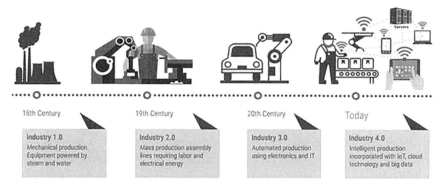

Fig. 4.1 Industrial revolutions summary.

mechanization of production through the use of steam power and the development of factory systems. The Second Industrial Revolution, took place in the late 19th and early 20th centuries, saw the rise of electricity, mass production, and the advent of the assembly line. The Third Industrial Revolution emerged in the mid-20th century with the introduction of computers, automation, and the Internet. Finally, we are now witnessing the onset of the Fourth Industrial Revolution, often referred to as the New Age of Digital, which encompasses advancements in AI, robotics, nanotechnology, and the IoT (Fig. 4.1).

This era of technological progress builds upon the previous three industrial revolutions and brings forth unprecedented connectivity, data-driven decision-making, and the integration of digital technologies into various aspects of our lives, paving the way for further innovation and transformation. It is a historic moment characterized by the rapid integration of digital technologies, such as AI, the IoT, big data analytics, and robotics into all facets of society and industry. It is the result of rapid advancements in digital technology and the increasing interconnectivity of physical and digital systems.

The widespread availability of high-speed Internet, cloud computing, mobile devices, and low-code applications has created the foundation for the digitalization of business processes and the creation of new, digital-based products and services. It is having a profound impact on the way work is done, leading to increased efficiency, reduced costs, and new business opportunities across all sectors.

In the energy sector, the digital revolution is transforming the way that energy companies operate, and energy services are being provided, enabling companies to collect and analyze vast amounts of data, providing insights into their operations, energy usage patterns, identifying opportunities for efficiency improvements, and helping to prevent equipment failures. It is likely to have far-reaching impacts on society and the global economy for generations to come. As new technologies and innovations emerge, organizations must stay ahead of the curve by embracing digital transformation and adapting to the changing technological landscape.

The new age of digital sets itself apart from previous industrial revolutions due to its unparalleled speed. While previous transformations took decades to unfold, this one moves at an extraordinary pace. Technological breakthroughs occur in rapid succession, continually pushing the boundaries of what is possible. The transformative power of this revolution is evident in the way it has revolutionized communication, commerce, entertainment, and countless other aspects of our daily lives. This convergence of technology is causing increased interconnectivity between physical and digital systems, leading to new levels of collaboration and innovation, and transforming the way work is performed and services are delivered. The traditional energy landscape is being disrupted and energy companies must recognize and adapt to this paradigm shift.

A digital transformation strategy within many energy companies is necessary for companies to remain relevant and stay competitive, to drive innovation, and capitalize on emerging opportunities. It is no longer a luxury but a necessity. The risks of falling behind in the rapidly evolving energy landscape can be catastrophic. Companies that fail to embrace digitalization risk losing market share, facing operational inefficiencies, and being unable to meet evolving customer demands, or worse.

At least 40% of all businesses will die in the next 10 years... if they don't figure out how to change their entire company to accommodate new technologies.

In the rapidly evolving landscape of the digital age, this quote by John Chambers, the former CEO of Cisco, carries profound implications for energy companies seeking to embark on a successful digital transformation journey. With technology advancing at an unprecedented pace, businesses in the energy sector face a critical imperative: adapt or risk obsolescence. Chambers' statement underscores the urgency for energy companies to embrace and integrate new technologies throughout their entire organizational structure. This entails not only implementing advanced digital tools but also fundamentally reimagining business models, processes, and strategies. To thrive in the face of increasing competition and changing customer expectations, energy companies must harness the power of digitalization to optimize their operations and continue to deliver value-added services. Failure to proactively embrace digital transformation may lead to their demise within the next decade, highlighting the need for decisive action and a holistic approach to embracing new technologies in the energy sector.

Digital maturity

The oil and gas industry faces unique challenges when it comes to digitization. It is not a straightforward process. Oil and gas companies are typically project-oriented and prioritize predictability of outcomes, which makes it challenging to integrate agile techniques into their organizational culture. While automation and data processing have been utilized in the industry for decades, many oil and gas companies mistakenly believe they are already proficient in digital and consider themselves digitally advanced. Paradoxically, the sector has been far slower than others in adopting digital solutions.

Furthermore, the industry's structure presents additional hurdles. Oil and gas companies heavily rely on oilfield services and engineering, procurement, and construction firms to carry out essential activities, creating interdependencies and making it difficult to achieve a consensus for change. Decentralization and mergers and acquisitions among operators have further complicated matters, resulting in localized management structures and a variety of legacy systems, which pose challenges in transitioning to digital solutions.

The "Digitalization Level" in the oil and gas sector refers to the extent of the adoption of digital technology by companies in this sector. Although the industry has been slow to adopt digital technology in the past, there is a growing realization of the necessity to embrace digitalization to stay competitive and keep up with changing customer needs.

The digitalization level varies greatly between companies and nations, with some companies making significant investments in digital technology and leading the way in digital transformation, while others are still in the early stages of incorporating digital solutions. Companies with a high degree of digitalization are better equipped to reduce costs, increase efficiency, and respond to changing customer demands. They are also more able to adapt to market changes and stay ahead of their competitors.

What is digital maturity?

Digital maturity is the measure of an organization's ability to create value through the deployment of digital technologies. It is also a key predictor of success for companies launching a digital transformation strategy. Businesses with high levels of digital maturity have a competitive advantage along multiple performance indicators, including revenue growth, time to market, cost efficiency, product quality, and customer satisfaction. Businesses with low levels of digital maturity struggle to achieve these benefits. Given digital's continued contribution to company performance, the gap between digital leaders and laggards will likely grow.

There are numerous indices that attempt to measure "how digital" companies are. The *Digital Maturity Index* is a measurement tool used to assess and quantify an organization's level of digital transformation and technological advancement. It evaluates how effectively a company utilizes digital technologies, processes, and strategies to achieve its business goals and remain competitive in the market. The index typically considers various factors, such as digital infrastructure, data management, employee digital literacy, customer engagement, innovation, and overall digital strategy. By measuring an organization's digital maturity, companies can identify their strengths and weaknesses, make informed decisions to optimize their digital efforts, and prioritize investments to drive further growth and efficiency.

In the context of the oil and gas industry, the Digital Maturity Index can play a crucial role in helping companies navigate the complex digital challenges in the sector. The industry is traditionally known for its reliance on conventional methods, but the increasing digitization of processes and technologies presents new opportunities. By using the Digital Maturity Index, companies can evaluate their current level of digitalization across their businesses, including exploration, production, refining, and distribution processes for operators and general services for contracting companies. This evaluation can reveal areas where technology can enhance efficiency, safety, and sustainability, such as implementing IoT sensors for remote monitoring, utilizing data analytics for predictive maintenance, or employing AI for project performance optimization. Harnessing digital tools effectively enable companies to streamline operations, reduce costs, minimize environmental impact, and stay competitive in an ever-evolving global energy landscape.

A Digital Maturity Index is a tool that evaluates the progress of organizations in their digital transformation journey. This assessment focuses on a series of dimensions that could include strategy, culture, technology, operations, and data. Measuring an organization's level of investment in digital technologies, integration of digital solutions into operations, and utilization of data for driving business outcomes enables the index to provide a numerical score that ranks organizations on a scale. This score offers a benchmark for organizations to gauge their digital progress, compare themselves with peers in their industry, and obtain a roadmap for achieving digital maturity and business success.

Embracing digitalization

Digital transformation has completely changed the working world. A constellation of innovations, ranging from cloud-based software to the IoT and automation, has propelled global businesses into an era of unparalleled connectivity, rapid problem-solving, and expanded market reach. By harnessing these digital technologies, companies can streamline processes, automate tasks, and leverage data-driven insights to drive innovation and achieve competitive advantage. Embracing digitalization opens doors to new markets, enables personalized customer experiences, and fosters collaboration and connectivity on a global scale. The strategic integration of these digital technologies offers companies the means to streamline their intricate processes, automate routine tasks, and harness the power of data-driven insights to catalyze innovation and establish a resounding competitive edge. Organizations that embrace digital solutions have greater resilience and a significant advantage on the competition.

At the heart of this paradigm shift lies the concept of embracing digitalization—a concept that transcends mere technological adoption to become a cornerstone of business evolution. The remarkable implications of this shift are manifold and resonant, painting a vivid picture of a future where companies can explore uncharted territories of success. By capitalizing on the endless arsenal of digital tools at their disposal, energy companies can drive efficiency into all aspects of their operations. Traditional manual processes give way to digital technologies that not only accelerate operations but also minimize wastage. This transition reclaims invaluable time, allowing personnel to divert their efforts toward revenue-generating activities and in doing so, strengthening the foundations of the organization.

However, the advantages of embracing digitalization extend far beyond mere efficiency gains. A resurgence of productivity emerges as businesses infuse their core business activities with collaboration technology and tools. The ability to go beyond geographical boundaries and cultivate a harmonious company culture takes center stage. The result is a dynamic and interconnected workforce that thrives on shared knowledge, collective creativity, and synchronous progress.

Security, an ever-pressing concern in the digital age, receives a heightened focus within the folds of digitalization. Organizations that embrace digital, fortified by digital defenses become more robust against the rising tide of cyber threats. By preemptively adapting to evolving security challenges, organizations can stand poised to better defend and respond to cyber threats with confidence.

Agility, which empowers organizations to navigate the unpredictable shifts in the energy industry landscape, finds a strong companion in digitalization. Equipped with insights drawn from data, businesses can not only make quicker decisions but also carry them out with exceptional speed. This capability to respond precisely when faced with change has the potential to turn a challenging situation into a chance for growth, placing a business at the forefront of advancement. Imagine a scenario where external circumstances unexpectedly shift, throwing conventional plans into disarray—this is where agility truly shines. Armed with the insights acquired from data analysis, organizations can swiftly analyze the situation, assess potential courses of action, and determine the most appropriate path forward. This agility in decision-making allows organizations to navigate obstacles and seize opportunities with a sense of precision that can turn adversity into advantage.

Moreover, it is not just about making quick decisions; it is about executing those decisions promptly and effectively. Digitalization provides the means to swiftly allocate resources, realign strategies, and mobilize teams, all of which are essential for converting decisions into tangible outcomes. This ability to translate decisions into action with swiftness and precision can be the turning point that transforms a crisis into an avenue for growth. In essence, the marriage of agility and digitalization empowers organizations to proactively adapt, respond, and capitalize on change. It positions them at the forefront of progress, enabling them to weather storms and emerge stronger and more innovative. As the business landscape continues to evolve at an ever-accelerating pace, the partnership between agility and digitalization becomes an indispensable asset, allowing organizations to forge a path to success amidst uncertainty.

In essence, the embracement of digitalization is a pivotal juncture that beckons organizations to seize their fate with both hands. It is a journey laden with transformative advantages—an expedition toward operational efficiency, heightened productivity, fortified security, and unparalleled agility. The digital age is upon us, and those who embrace it are poised to move forward on a journey of unparalleled growth and innovation.

Digital engineering

Engineering continues to play a fundamental role in driving the global economy, and in today's world, engineers are embracing a more collaborative approach and gaining access to an unprecedented wealth of data. The advent of digitization has revolutionized the way engineers conduct their work and make crucial decisions, opening up a vast array of technological possibilities. Digital transformation has brought about significant advancements in engineering, a field that has always been known for its rapid evolution. The integration of new technologies has accelerated product development, streamlined transportation processes, and facilitated real-time data management. Consequently, engineering has become an industry driven by data, making it imperative for businesses to adopt the right technologies to ensure sustainable growth within the sector.

Cloud computing stands out as one of the primary ways through which digitalization has profoundly impacted the engineering landscape. Although not a novel concept, cloud computing has evolved to become a more effective and efficient tool for engineers. By leveraging cloud-based platforms, engineers can now store their digital designs in a centralized location, streamlining collaboration and project management before physical construction even begins. The true value of cloud computing in engineering lies in its capabilities for data capture, 3D modeling, and creating enhanced customer experiences. As the engineering industry grows increasingly data-driven, cloud computing has become an indispensable asset for professionals seeking to optimize their processes and improve their outcomes.

Digital twins

In the wake of the wave of digital transformation and the expansion of big data, engineering firms have embraced a groundbreaking concept known as digital twins. This concept has yielded profound outcomes that are redefining the approach to design, analysis, and maintenance for energy companies.

Digital twins, at their core, represent an ingenious fusion of the virtual and physical spheres. In practice, this involves creating intricate virtual replicas of tangible objects, a feat enabled by cutting-edge technologies. These digital twins become dynamic mirrors of real-world assets, endowed with the capacity to not only mimic their form but also emulate their behavior and characteristics. The marriage of technology and innovation is not merely an intellectual pursuit; it holds tangible benefits that ripple through the core of engineering operations. Imagine engineers standing at the crossroads of design, grappling with the intricate details of a complex system. The advent of digital twins provides them with an invaluable tool, granting them the ability to create a virtual counterpart of their creation. This virtual sibling, meticulously woven from lines of code, stands ready to undergo tests, simulations, and analyses, all within the ethereal confines of a computer's memory.

The significance of this becomes most apparent when predictive analysis comes into play. Armed with the digital twin, engineers can delve into a realm of foreseeability that was previously reserved for science fiction. They can orchestrate simulations under a multitude of conditions, unleashing the power to witness firsthand how their physical creation will perform in various scenarios. It is akin to peering into a crystal ball that reveals not vague prophecies but concrete and data-backed insights.

Beyond the scope of predictions, digital twins also lay the foundation for proactive strategies in maintenance. Engineers can employ these virtual companions to forecast the wear and tear of physical assets, foreseeing maintenance needs even before the first signs of decay emerge. This predictive prowess flips the conventional reactive approach on its head, as engineers now hold the keys to preventing issues rather than resolving them.

In effect, the advent of digital twins marks a noteworthy milestone in the history of engineering innovation. This concept is redefining the very contours of how products and systems are conceived, nurtured, and maintained. It is a journey that spans the borders of both the tangible and the intangible, where the digital twin bridges the gap between concept and reality. As engineering firms continue to chart their course in this digital age, the digital twin stands as a testament to their capacity to blend vision and technology, forever transforming the way we interact with the physical world.

One compelling example of a digital twin's application in the energy sector can be found in a modern power plant. Let us take a look at a natural gas power plant and explore how the implementation of a digital twin has brought about groundbreaking benefits. In this scenario, engineers have created a digital twin of the entire power plant, including its various components such as gas turbines, boilers, generators, and cooling systems. This digital twin is a highly detailed virtual replica, capturing not only the physical dimensions of the equipment but also their behaviors, interactions, and responses to changing conditions.

Imagine that the power plant needs to undergo a scheduled maintenance shutdown. Traditionally, such shutdowns are planned based on pre-determined intervals or best estimates of equipment wear and tear. However, with the digital twin in place, engineers can take a more data-driven and precise approach. They feed real-time operational data from the physical plant into the digital twin, allowing it to continuously update and refine its under-standing of the plant's condition.

As the digital twin analyzes the data, it can simulate how the various components will perform in the near future. It takes into account factors such as temperature, pressure, stress, and operational load. By running these simulations, the digital twin can accurately predict the deterioration of spe-cific parts and pinpoint when they are likely to require maintenance.

The groundbreaking benefits of this approach become evident. Rather than conducting maintenance based on broad assumptions, the power plant can now optimize its shutdown schedule. This not only minimizes down-time but also prevents unnecessary maintenance, conserving resources and reducing costs. Additionally, the plant can avoid unexpected breakdowns that might have occurred without the insights provided by the digital twin.

Furthermore, the digital twin can be used to experiment with different operational scenarios. Engineers can test the impact of adjusting certain parameters, such as fuel mix or load distribution, to optimize efficiency and emissions. This virtual testing minimizes the need for risky or time-consuming changes in the physical plant.

In summary, the digital twin in the energy sector, as illustrated by the power plant scenario, offers unparalleled advantages. Its ability to predict equipment behavior, optimize maintenance schedules, and simulate various operational scenarios transforms the conventional energy infrastructure into a highly efficient, cost-effective, and responsive system. It not only drives operational excellence but also sets the stage for innovation and sustainable energy practices in the ever-evolving energy landscape.

Engineering collaboration

The integration of digital technology has ushered in a profound and trans-formative change in the way engineers collaborate and innovate. The emer-gence of digitization has dismantled traditional barriers, reshaping the foundations upon which engineers collaborate and bring their inventive visions to life. This evolution has sparked a renaissance of connectivity, where engineers no longer remain confined by geographic limitations, but instead traverse the globe to cultivate a new era of collective ingenuity.

In this era of technological synergy, the once-familiar landscape of collaboration has undergone a transformation. Engineers now find themselves operating within an expansive realm of possibilities, where boundaries are dissolved by the power of digital tools. Virtual meetings, once a rarity, have blossomed into a pivotal means that exceeds the constraints of distance and geography. Through these digital rendezvous, minds converge from distant corners of the world, fostering a cross-pollination of ideas and perspectives that enriches the tapestry of problem-solving. Central to this transformation is the ascent of cloud-based collaboration tools that serve as digital workspaces where engineers converge to pool their talents. This virtual meeting room has become the epicenter of innovation, where insights, ideas, and data flow in a torrent of creativity. Real-time data sharing, a cornerstone of this digital renaissance, has not only expedited the exchange of vital information but has also unveiled new dimensions of collaboration, unbinding engineers from temporal constraints and imbuing their interactions with immediacy and agility.

With these digital avenues, engineers now stride forward in the pursuit of solutions that seemed insurmountable before with collective intellect. The collaborative energy cultivated in this digital era capitalizes on multidisciplinary expertise, transforming hurdles into steppingstones toward innovation.

But the transformation goes beyond mere acceleration of problem-solving. A profound cultural shift has taken root. The spirit of sharing, inherent to the engineering ethos, has found new expression in the digital era. As knowledge flows seamlessly through virtual conduits, barriers that once impeded the propagation of ideas are dismantled. This culture of boundless exchange nurtures an environment where innovation is not an isolated endeavor, but an ongoing collective endeavor nurtured by the synergy of diverse minds.

Digital transformation has ushered in a new era of possibilities for engineering firms. The integration of cloud computing has enhanced data management and project efficiency, while the advent of digital twins has enabled predictive analysis and optimization of real-world assets. Moreover, the collaborative nature of engineering has been revolutionized, as engineers from different corners of the world can now work together seamlessly, capitalizing on the wealth of available data to drive innovation and sustainable growth in the sector. As technology continues to evolve, engineers must remain adaptable and open to embracing new digital tools and strategies to stay at the forefront of this ever-evolving industry.

The future of engineering?

The field of engineering stands at the precipice of a transformative era, poised to redefine its methods, scope, and impact on society. As the world undergoes rapid technological advancements and shifts toward sustainable practices, the future of engineering is bound to be marked by complexity, innovation, and digital transformation. This evolution is set to be driven by the integration of AI and the establishment of a robust digital thread that guides projects from concept to handover and operations.

The challenges of the future, especially those linked to the energy transition, demand a new breed of engineers who are adept at navigating uncertainty and complexity. Climate change, a persistent global issue, has placed sustainable practices at the forefront of engineering endeavors. This compels engineers to design structures that not only fulfill their functional requirements but also possess a positive environmental impact. As the drive toward renewable energy sources, smart infrastructure, and resource efficiency gains momentum, engineers will need to think beyond conventional solutions and embrace interdisciplinary collaboration to create holistic, eco-friendly solutions.

AI is set to revolutionize the engineering landscape. Through advanced machine learning algorithms, AI can process and analyze massive datasets, enabling engineers to make more informed decisions. This is particularly significant in design optimization, where AI can rapidly generate and evaluate numerous design options, leading to innovative solutions that might not have been feasible through traditional methods. AI-powered predictive analytics can enhance project management by identifying potential issues before they escalate, enabling proactive interventions. Robotics and automation, fueled by AI, are expected to play a pivotal role in construction and maintenance tasks, enhancing efficiency, safety, and precision.

The digital thread

The concept of the digital thread has emerged as a transformative force in the advancement of engineering as a function. This visionary approach involves the creation of a single, integrated data source that weaves its way through the entire lifecycle of a project, from its very inception as a 3D model, through the stages of procurement, construction, and even ongoing maintenance and operations. The digital thread is rapidly becoming a cornerstone of modern engineering practices, and its implications are nothing short of revolutionary.

At its core, the digital thread serves as an information superhighway, ensuring that all stakeholders involved in a project are equipped with access to real-time, accurate data. Gone are the days of disjointed processes and information silos that hindered collaboration and introduced errors into projects. With the digital thread in place, engineers, architects, suppliers, and construction teams are all connected seamlessly, fostering a culture of collaboration and transparency. One of the primary advantages of the digital thread is its ability to reduce errors and streamline decision-making processes. In the past, miscommunications and outdated information often led to costly mistakes and project delays. With the digital thread, these pitfalls are minimized, if not eliminated altogether. Real-time data updates and a shared digital environment ensure that everyone involved is working with the most up-to-date information. This not only accelerates project timelines but also enhances the overall quality of the end product.

One of the most exciting extensions is the incorporation of augmented reality (AR). AR technology overlays digital information onto the physical world, creating an immersive and interactive experience. Within the context of the digital thread, AR has found a multitude of applications. It can provide real-time guidance for tasks like assembly, training, and quality control.

Imagine a construction worker wearing AR glasses that display step-by-step instructions for assembling a complex structure, all while offering real-time feedback on their progress. This not only expedites construction processes but also enhances safety and reduces the learning curve for new workers. Additionally, AR can be utilized for quality control inspections, allowing inspectors to view digital overlays that highlight potential issues or areas that require attention.

The digital thread is ushering in a new era of engineering development by revolutionizing the way information flows through a project's lifecycle. Its ability to connect stakeholders, reduce errors, and facilitate the integration of cutting-edge technologies like AR make it an indispensable tool for modern engineering endeavors. As we continue to push the boundaries of what is possible in engineering, the digital thread stands as a testament to the power of seamless information integration in achieving unparalleled levels of efficiency, collaboration, and innovation.

AI and ML

Even though AI has its roots in academic and scientific pursuits dating back to the 1950s, the technology has experienced a remarkable surge in popularity and practical application in recent years. This surge is unparalleled in the history of AI, with unprecedented levels of research, investment, and real-world business implementations. A report by Grand View Research predicts that by 2030, the global AI market size is expected to skyrocket to $1811.8 billion, a significant leap from the $136.6 billion recorded in 2022, boasting an impressive compound annual growth rate of 38.1% [1].

AI is not merely a technological phenomenon; it carries the promise of transforming not just businesses but also enhancing the overall human experience. Before we delve deeper into the implications and impacts of AI, let us establish a foundational understanding of what AI truly represents.

John McCarthy, an American computer scientist and cognitive scientist, is credited with coining the term "artificial intelligence" in 1956. He offered a concise yet profound definition, describing AI as "the science and engineering of making intelligent machines." This definition encapsulates the core mission of AI, which is to create machines that can mimic and perform tasks associated with human intelligence [2]. Expanding on this, AI, as defined by Science Daily, is "the study and design of intelligent agents where an intelligent agent is a system that perceives its environment and takes actions that maximize its chances of success." This definition underscores the pivotal role of AI in creating systems capable of not only perceiving their surroundings but also making informed decisions to achieve their objectives.

In simpler terms, AI grants computers the ability to perform tasks traditionally associated with intelligent beings, particularly humans. This goes beyond basic automation; AI endows these systems with the capacity to reason, interpret meaning from data, and learn from their experiences. In essence, AI strives to instill human-like cognitive capabilities into machines, enabling them to adapt, evolve, and excel in a wide range of applications, from solving complex problems to assisting with everyday tasks.

As AI continues to evolve and integrate into various aspects of our lives, its impact extends far beyond the realm of technology, making it a transformative force in the 21st century. It is poised to reshape industries, improve efficiency, and enhance our interaction with technology, ultimately redefining the way we work, live, and relate to the world around us.

One domain where the impact of AI is particularly profound is in engineering and project management, especially in the context of the ongoing energy transition. The energy sector is undergoing a significant transformation as it seeks to reduce carbon emissions and transition to sustainable and renewable sources of energy. AI is a powerful tool in this endeavor, offering solutions that enhance the efficiency and sustainability of energy projects.

In the field of engineering, AI is driving a profound revolution in the design and operation of energy systems, particularly in the context of renewable energy sources. One of the key strengths of AI in this domain lies in its ability to analyze immense datasets with speed and precision, offering solutions that were previously challenging, if not impossible, to attain. AI-powered algorithms are instrumental in optimizing the design and operation of renewable energy facilities, such as wind farms and solar arrays. These algorithms excel at sifting through a vast array of geospatial and meteorological data to determine the ideal locations and configurations for these energy installations. By considering factors, such as wind patterns, solar irradiance, geographical features, and even local energy demand, AI can pinpoint the most suitable sites for renewable energy projects. This optimization not only increases energy output but also maximizes the return on investment, making the transition to clean energy sources more economically viable.

Furthermore, AI does not stop at the planning phase; it extends its benefits to the operational aspect of energy infrastructure. AI-driven predictive maintenance models are instrumental in identifying equipment maintenance needs before they lead to costly breakdowns. These models analyze real-time data from sensors and historical performance data to detect signs of wear and tear, enabling proactive maintenance. This not only reduces downtime but also extends the lifespan of critical equipment, thus resulting in substantial cost savings and increased reliability in energy production.

In addition to these benefits, AI plays a pivotal role in energy infrastructure performance enhancement. By continuously monitoring and analyzing data from energy systems, AI algorithms can identify inefficiencies and recommend adjustments in real-time. This ongoing optimization ensures that energy infrastructure operates at peak efficiency, minimizing energy waste and environmental impact.

Moreover, AI-driven simulations provide engineers with a powerful tool for testing various scenarios and fine-tuning their designs. When applied to complex energy projects, simulations can model a range of conditions, from weather fluctuations to energy demand variations. Engineers can use these simulations to assess the resilience and performance of their designs under different circumstances, allowing them to make data-driven decisions and adapt their projects as needed.

The integration of AI in engineering and energy system design is not only optimizing the transition to sustainable energy but is also making it more intelligent, efficient, and economically viable. This synergy between AI and engineering is paving the way for a greener and more sustainable future, where clean energy sources are harnessed to their full potential, reducing environmental impact and providing more reliable energy to meet the growing demands of our modern world.

In project management, AI assists in streamlining and optimizing the execution of energy projects. AI-powered project management tools can predict potential delays and bottlenecks, helping project managers allocate resources effectively and ensure timely completion. Analyzing historical project data allows AI to improve cost estimations and reduce budget over-runs. Additionally, AI-enhanced risk analysis can aid in identifying and mit-igating potential project risks, making energy projects more predictable and reliable.

Overall, AI is poised to play a pivotal role in engineering and project management within the energy transition. It is already accelerating the adop-tion of sustainable energy solutions, as well as enhancing the efficiency and sustainability of energy projects, contributing to a greener and more envi-ronmentally conscious future. As AI continues to evolve and intertwine with these domains, its transformative impact promises to be a driving force in achieving the goals of the global energy transition.

Case study: ADNOC Partner to accelerate AI solutions

In a landmark move, ADNOC, a leading energy company, has par-tnered with AI and Big Data Analytics leader Presight to drive AI integration in the energy sector. This strategic partnership, announced on May 1, 2024, is set to accelerate the adoption and development of AI solutions, leveraging Presight's advanced capabilities and ADNOC's extensive industry expertise.

The partnership combines AIQ's breakthrough AI energy solutions with Presight's cross-sector big data analytics, product development, and interna-tional market access. This integration is expected to realize operational efficiencies and synergies while meeting a wider array of customer require-ments. By combining their capabilities, ADNOC and Presight aim to deliver greater value for themselves as well as for the UAE, and the global energy sector.

ADNOC has long been focused on becoming the most AI-enabled energy company globally, leveraging AI to drive productivity, economic growth, and environmental sustainability. For decades, ADNOC has devel-oped its digital infrastructure, enabling rapid business expansion, unlocking untapped opportunities, and responsibly supplying energy to meet customer needs. In 2023, ADNOC announced that it generated $500 million in value by deploying AI solutions across its value chain. These AI applications, integrated into field operations and corporate decision-making processes, have also contributed to significant environmental benefits by abating up to 1 million tons of CO_2 emissions, equivalent to removing around 200,000 gasoline-powered cars from the road.

The partnership with Presight is expected to further enhance AIQ's capabilities. AIQ, launched in 2020, has quickly established itself as a leading provider of AI solutions in the energy sector, reinforcing Abu Dhabi's position as a leading global hub for AI. AIQ has already developed over 20 AI applications and filed 16 patents. These key AI tools are driving efficiency and unlocking value across ADNOC's operations by identifying, monitoring, and reducing emissions, and improving safety. The AI solutions, trained using vast amounts of current and historical data, are being commercialized and will form the foundation for the transformative growth of AIQ.

The strategic significance of this partnership cannot be overstated. Integrating AIQ and Presight's big data analytics and AI offerings enables ADNOC to enhance its operational efficiencies and position itself to meet a wider array of customer requirements and deliver greater value. This collaboration underscores ADNOC's commitment to pioneering AI technology in the energy sector, reinforcing Abu Dhabi's position as a global AI hub.

The future outlook for this collaboration is promising. H.E. Mansoor Al Mansoori, Chairman of Presight, highlighted that since its debut on the Abu Dhabi Securities Exchange (ADX) in 2023, Presight has showcased robust growth, sustainable profitability, and an ability to attract top-tier talent and global clientele. This transaction is expected to unlock both Presight and AIQ's immense growth potential, positioning the two organizations for unprecedented local and global success. Presight's acquisition of AIQ, with plans to serve ADNOC and the wider energy sector, further reinforces their commitment to the wellbeing of the planet and its inhabitants.

ADNOC's strategic partnership with Presight exemplifies the imperative for energy companies to embrace digital transformation. Integrating advanced AI solutions is allowing ADNOC to set a benchmark for leveraging technology to drive efficiency, sustainability, and value in the energy sector. This case study highlights the critical role of digital adoption in navigating the future of energy, demonstrating how embracing digital is not just a strategic advantage but a necessity for staying competitive and responsible in the modern energy landscape [3].

Example: AI-driven trouble shooting

With soaring feedstock costs, energy companies are more concerned than ever about the possibility of an unexpected downtime on their assets. Despite strides in industrial technology, the foundational principles of instrumentation and monitoring practices have remained largely unchanged for half a century. We design our plants with interlocks that trip the plant to preserve safety, and rules-based alarms that alert operators, however, still have surprises, which highlights the need for a more proactive approach. Monitoring solutions emerge as a beacon of hope, empowering teams to monitor system states in real-time by gathering predefined metrics and logs. However, the true game-changer arrives with AI-assisted troubleshooting platforms, ushering in a new era of predictive maintenance by uncovering unforeseen patterns and anomalies. One of the key value-add case studies for AI is in real-time troubleshooting of equipment and assets, mirroring the transformative impact witnessed in IT operations. By bridging the gap between traditional monitoring and AI-driven diagnostics in physical infrastructure, operators can mitigate surprises, prevent downtime, and optimize operational efficiency.

The integration of AI-driven troubleshooting platforms signifies a paradigm shift in energy production processes, offering operators unprecedented insights into system anomalies and potential breakdowns. Strategically deployed during shift changes, plant-wide meetings, and continuous monitoring in control rooms, these platforms enable proactive problem-solving and enhance cross-departmental collaboration. Illuminating specific subsystems and sensor anomalies in real-time enables operators to direct their attention to critical areas before issues escalate. This proactive approach mitigates the risk of unforeseen downtime and minimizes environmental repercussions such as flaring, thereby bolstering operational resilience and sustainability.

Embracing the technological evolution of AI-driven troubleshooting platforms exceeds mere risk mitigation; it cultivates a culture of proactive problem-solving and continuous improvement. These platforms enhance overall operational efficiency and visibility and in doing so, foster synergy among operations, engineering, and reliability teams.

Through seamless integration into daily operations, from shift changes to round-the-clock monitoring, AI-driven solutions can empower operators to stay ahead of the curve, anticipate challenges, and optimize performance.

As energy producers navigate the complexities of an ever-evolving landscape, embracing AI-assisted troubleshooting represents a strategic imperative in safeguarding against surprises, maximizing uptime, and driving sustainable growth.

Digital transformation for energy transition

Recognizing the immense potential that digitalization strategies hold for driving the energy transition, many international oil companies (IOCs) have already embarked on their digital journeys, adopting digital technologies and systems to enhance their businesses. Shell has invested significantly in cutting-edge digital technologies, such as advanced data analytics and AI. These investments are not merely for the sake of technological advancement, but rather with a clear purpose: to optimize operations, enhance safety measures, and significantly reduce their environmental footprint.

Shell's digitalization initiatives have fundamentally reshaped the way they operate, making data the bedrock of their decision-making processes. By collecting and analyzing vast amounts of data from their operations, they have gained valuable insights that inform their strategies and drive their actions. This data-driven approach empowers Shell to optimize their processes, minimize downtime, and enhance operational efficiency across their diverse range of energy-related activities. From the exploration and production of hydrocarbons to the development of renewable energy solutions, digitalization has enabled Shell to make informed choices that align with their vision for a sustainable future.

In their offshore oil and gas production operations, Shell utilizes real-time data collected from sensors installed on production platforms. By analyzing these data, they gain insights into equipment performance, production rates, and potential maintenance needs. This allows them to make proactive decisions, reducing unplanned downtime and optimizing production efficiency. The application of advanced data analytics enhances operational performance and also contributes to cost savings and improved safety measures.

Robotics and automation technologies is another key area that IOCs have embraced to enhance performance in their operations. In their refineries and chemical plants, Shell employs robots equipped with advanced sensing capabilities to perform inspections in hazardous areas. These robots can navigate through hazardous environments, collect data, and provide accurate assessments of equipment conditions. By deploying robots for inspections, Shell reduces the exposure of workers to potentially dangerous situations, improving overall safety standards. The integration of robotics not only enhances safety but also increases operational efficiency by reducing inspection time and enabling more effective maintenance planning.

Spot, the robotic creation of Boston Dynamics, holds immense potential to drive the digital revolution within the energy sector, offering invaluable support to companies like Shell. This quadrupedal robot boasts exceptional agility and adaptability, making it a groundbreaking technology with wide-reaching applications.

With its autonomous navigation capabilities and an array of sensors and cameras, it can traverse complex and potentially hazardous environments, such as oil rigs, refineries, and power plants. During these inspections, Spot can swiftly identify and report anomalies, potential leaks, or equipment malfunctions in real-time, significantly enhancing safety and operational efficiency.

Spot also plays a pivotal role in maintenance tasks. Its dexterous limbs enable it to conduct routine inspections and perform basic repairs, particularly in challenging environments where human intervention can be perilous. This capability can eliminate the need for human inspections in hazardous situations within high-hazard facilities. The 65-pound industrial robot dog uses "remote operation and autonomous sensing" to help with data collection and inspections. It is equipped with an array of sensors and can gather crucial data concerning factors like temperature, pressure, and gas emissions to support predictive maintenance, operational optimization, and compliance with environmental regulations. Leveraging Spot's capabilities enables energy companies to advance their digital transformation efforts, striving for more efficient, safer, and sustainable energy operations.

In Shell's Alberta refinery, it has invested two Boston Dynamics robot dogs. Shell's has put the dogs to work, tacking "riskier" and "mundane" tasks at the plant. "They're going to start doing inspections in our plants," said Conal MacMillan with Shell Scotford. "They'll be able to 3D image some of our units, they'll be able to do some thermal scanning." [4] (Fig. 4.2).

Shell's embrace of digital transformation has empowered them to optimize their operations and improve safety but has also driven them toward a low-carbon energy future. Investing in advanced data analytics, AI, and robotics has enabled Shell to pave the way for data-driven decision-making, operational efficiency, and accelerated progress in the transition to renewable energy sources [5].

Fig. 4.2 Boston Dynamics robot dog.

In the energy transition era, digitalization is not just an option; it is a vital strategy for energy companies to remain relevant and competitive. By embracing the digital revolution, deploying fit-for-purpose digital solutions, and fostering a culture of innovation, energy companies can unlock new opportunities, improve operational efficiency, and stay ahead in the rapidly evolving energy landscape.

Example: Transforming the customer experience

The integration of advanced digital technologies is revolutionizing the way customers interact with energy service providers, ushering in a new era of enhanced customer experiences. Through the use of online portals and mobile apps, energy companies are empowering customers with real-time information and services that were previously inaccessible. For instance, customers can now easily monitor their energy usage, receive detailed billing information, and access personalized energy management tools at their fingertips. This digital transformation simplifies the customer experience and also enables them to make more informed decisions about their energy consumption, leading to increased efficiency and cost savings.

Moreover, the decentralization of energy production and distribution is another key aspect of the digital transformation in the energy sector, contributing to the transformation of the customer experience. Technologies like microgrids and rooftop solar panels enable customers to generate their own energy and even sell excess energy back to the grid. This newfound energy independence grants customers greater control over their energy consumption and allows them to actively participate in the energy transition. Energy service providers are capitalizing on this trend by offering tailored solutions to customers, such as energy storage systems that optimize energy usage, energy management systems that enable intelligent control of devices, and energy efficiency services that help customers reduce their carbon footprint. These customized energy solutions cater to the unique needs and preferences of individual customers, fostering stronger relationships and increasing customer satisfaction.

Real-world examples of these digital innovations in the energy sector abound. For instance, companies like E.ON, a major European energy provider, have developed mobile apps that allow customers to monitor and control their energy usage remotely. Through these apps, customers can adjust heating and cooling settings, track energy consumption in real-time, and receive personalized energy-saving tips.

Another example is the German energy company Sonnen, which offers integrated energy management systems that combine solar panels, battery storage, and intelligent software. These systems enable customers to generate, store, and manage their own renewable energy, leading to significant cost savings and reduced reliance on the grid. These examples demonstrate how digitization is not only transforming the customer experience but also empowering customers to actively participate in the energy transition and contribute to a more sustainable future.

How "digital" is your organization?

Just "doing" digital things will not make an organization more digital. While some companies have explored the benefits of digitalization and "dabbled with digital," others have moved toward business, operating and customer models that are optimized for digital and are profoundly different from prior siloed and unsynchronized models.

Digital maturity frameworks are essential tools that help organizations assess their current level of digital proficiency, establishing a baseline that guides their journey toward developing stronger digital capabilities. These frameworks clarify the stages of digital maturity, allowing organizations to benchmark their digital capability and proficiency against industry standards, identify gaps, and prioritize improvement efforts. They also provide a structured roadmap for digital transformation, outlining clear steps and milestones to progress through higher levels of digital maturity.

Through these frameworks, companies can seek alignment and collaboration across departments, ensuring a shared understanding of digital goals and coordinated efforts toward transformation. Additionally, digital maturity models promote continuous improvement by encouraging regular assessment and refinement of digital strategies, enabling organizations to remain competitive and resilient in a rapidly evolving technological landscape.

There are many digital maturity frameworks available today, each offering unique perspectives and criteria to assess and guide an organization's digital transformation. These models vary widely in structure and focus areas, allowing organizations to choose one that aligns closely with their specific needs and industry context. Examples include frameworks developed by prominent institutions such as MIT CISR, which emphasizes strategy, organization, processes, and technology; Gartner, focusing on strategy, organization, operations, customer experience, and technology; and Deloitte, which incorporates strategy, organization, capabilities, insights, and customer experience. Models from McKinsey and KPMG also address digital maturity across multiple dimensions, including organization, customer experience, and technology, offering tailored approaches to foster innovation and continuous improvement.

To effectively leverage a digital maturity model, organizations should follow a structured, step-by-step approach. First, selecting the appropriate model is crucial, this choice should align with the organization's goals, industry requirements, and transformation ambitions. A comprehensive assessment is then conducted, involving stakeholders and experts to evaluate the organization's maturity level through data collection, interviews, and workshops. The results are then analyzed to identify strengths, weaknesses, and areas for improvement. From there, organizations can develop a detailed roadmap and action plan that includes specific goals, priorities, and resource allocations, ensuring alignment with both current needs and long-term objectives.

The final steps involve implementing the action plan with ongoing monitoring, using KPIs to track progress, and adapting to any changes in business needs or market conditions. Continuous improvement is essential, as digital transformation is an evolving journey. Periodic reassessing of digital maturity and refining strategies enables organizations to stay agile, respond effectively to technological advancements, and secure a competitive edge in an increasingly digital landscape. This structured approach ensures that digital maturity models guide transformation efforts as well as empower organizations to thrive amidst rapid change. A typically Digital Maturity Model or spectrum consists of a number of levels. An indicatory Digital Maturity Model is presented below to illustrate how it can add value in determining where an organization sites in terms of its digital maturity (Fig. 4.3).

Initiator

Organizations at the Initiator stage are just beginning their journey toward digital transformation. Digital efforts are limited and often exist in isolated pockets, such as individual departments experimenting with digital tools on a small scale. Leadership may not yet fully understand or prioritize digital initiatives, resulting in minimal alignment across the organization. Processes remain largely manual, and digitalization is not seen as a core component of the business strategy. At this stage, the organization lacks an enterprise-wide vision for digitization, and any efforts that do exist are fragmented and experimental.

Explorer

In the Explorer stage, organizations start to acknowledge the potential of digital transformation and take their first coordinated steps toward adopting digital tools. While digital initiatives begin to appear across multiple

Fig. 4.3 Indicatory Digital Maturity Model.

departments, they are typically unstructured and lack consistent execution. This stage is characterized by early adoption of basic digital technologies and limited integration within existing processes. Organizations in this stage face challenges in orchestrating their digital efforts and need a more strategic approach to align these initiatives with broader business goals. The groundwork for digitalization is still in a developmental phase, and there is a recognition that further efforts are needed to mature these practices.

Adopter

Organizations in the Adopter stage have embraced digital solutions in various parts of the business and gained moderate experience with digital tools and processes. Multiple departments now rely on digital practices to enhance efficiency and productivity, though these efforts may still lack a unifying strategy. At this stage, digital transformation is gaining momentum, but there is a need for cohesive leadership to drive a company-wide digital strategy. The organization may face challenges in ensuring consistency and alignment across departments, resulting in potential inefficiencies or duplication of efforts. Moving beyond this stage requires establishing strong leadership to champion digital initiatives and ensure that they are integrated into the organization's long-term vision.

Strategist

Organizations in the Strategist stage have recognized the strategic importance of digital transformation and are working to embed digital practices throughout the business. Digital adoption has become organization-wide, with efforts aligned to meet strategic goals and deliver tangible value. Departments are more collaborative in their approach, and digital processes are increasingly standardized. While the company has made significant strides in digital integration, there may still be areas that require further refinement to achieve seamless digital operations. This stage is characterized by a proactive approach to digital transformation, with a focus on continuous improvement and preparing the organization to move to the next level of maturity.

Innovator

At the Innovator stage, organizations have achieved a high level of digital maturity, with digital technologies and practices deeply integrated into every aspect of the business. These organizations actively leverage digital solutions to drive innovation, optimize operations, and gain a competitive edge in the market. Digital transformation is not just a project, it is embedded in the culture, with leadership fostering a mindset of continuous innovation and adaptability. Innovators are at the forefront of adopting new technologies and exploring ways to future-proof the organization. They embrace technology as a core enabler of business success and are well-positioned to respond to changes in the market, making them leaders in their industry.

The Digital Maturity Model provides a roadmap for companies as they embark on their digital transformation journey. From the initial stages of exploring digital possibilities to the advanced level of being truly digital, each phase represents a level of maturity and integration of digital capabilities. To move along the Digital Maturity Model, companies must address several challenges. These may include overcoming resistance to change, ensuring data security and privacy, cultivating a digital culture, and investing in employee upskilling. The successful navigation of these challenges can enable companies to unlock the full potential of digitalization and gain a competitive advantage in the digital era. As companies progress along the Digital Maturity Model, they unlock new opportunities for innovation, efficiency, and growth. Embracing digital transformation is not just a technological shift; it requires a strategic, cultural, and organizational transformation.

Digital citizen: Cultivating a culture of collaboration and innovation

Cultivating a culture of collaboration and innovation is paramount for energy companies embarking on their digital transformation journey. To bolster this, establishing a Centre of Excellence (CoE) can be instrumental. CoE can serve as a hub for best practices, skills enhancement, and cutting-edge solutions. Shifting toward a more innovative and agile culture is essential to fully embrace the opportunities that digitalization presents. It begins with encouraging employees at all levels to actively participate in the process of digital transformation. Companies can establish cross-functional teams, innovation labs, or even leverage the CoE to explore digital possibilities, fostering an environment where fresh ideas can flourish. By valuing experimentation and viewing failure as an opportunity for learning, energy companies empower their workforce to drive digital innovation. We will delve into the benefits a CoE can bring and how to setup a CoE in Chapter 6.

The concept of a "digital citizen" introduces a fundamental change in the way energy companies interact with their customers and stakeholders. In this context, a digital citizen is not just a passive recipient of services but an active participant in the digital ecosystem created by these companies. Digital citizens engage with energy providers through various digital channels and platforms, forming a dynamic and interconnected community. Embracing the digital citizen concept holds several key advantages for energy companies:

First, it opens doors to co-creation and collaboration. Energy companies can leverage the collective intelligence of their digital citizen community, involving them in the ideation and development processes. This approach taps into a diverse range of perspectives and expertise, enabling customers and stakeholders to provide feedback, suggest improvements, and even contribute to the design of new energy products and services. Collaborative platforms and crowdsourcing initiatives become essential tools for harnessing this collective creativity.

Second, by actively engaging with digital citizens, energy companies gain insights into evolving customer needs and preferences. This real-time feedback loop allows companies to adapt and tailor their offerings to meet the ever-changing demands of their digital-savvy audience. It helps in creating customer-centric solutions that enhance overall satisfaction and loyalty. Furthermore, the digital citizen concept fosters a sense of community and loyalty among customers. When individuals and entities feel that they have a stake in the energy company's digital ecosystem, they are more likely to remain engaged and committed. This can lead to long-term partnerships and a strengthened brand image.

Digital transformation roadmap

The energy sector is experiencing a paradigm shift as it transitions toward more sustainable and efficient energy solutions. This energy transition necessitates not only the adoption of new technologies but also a fundamental change in the way energy organizations operate. One of the pivotal changes required is for energy companies to embark upon their own digital transformations. Here is how energy companies can develop a digital transformation roadmap.

Accepting that digital transformation is an imperative

You can't delegate digital transformation for your company... You and your organization have to own it. Company leaders need to engage, embrace, and adopt new ways of working with the latest and emerging technologies.

Digital transformation is not just about adopting the latest tools or technologies, and it is about fundamentally altering how the organization functions. From operations to customer engagement, from resource management to strategic decision-making, every facet of the business needs to be reimagined through the lens of digital possibilities.

A key part of this step is to create a bold digital vision from the top. Creating a compelling digital vision that stems from leadership is essential. For the company's aspirations to resonate throughout the entity and motivate employees, it is imperative for the company's leadership to clearly define and embrace their vision for a digital future. Simply detailing how technologies play a role in daily operations will not suffice. A potent vision should clarify how digital strategies will aid in achieving overarching goals, be it reducing the company's carbon footprint, enhancing profitability, improving safety etc. This vision's formulation demands substantial participation from leaders across various sectors, ensuring its feasibility. Moreover, this vision must be audacious enough to serve as a guiding beacon for future investments and initiatives.

Assess your organization's digital maturity

To successfully navigate the intricate landscape of digital transformation, it is paramount for businesses to recognize their current position within the Digital Maturity Model. There are frameworks and tools, specifically designed to aid organizations in this assessment. Such Digital Maturity Models provide a comprehensive view, allowing companies to understand their progression on a continuum that stretches from "Digital Initiator" where digital initiatives are implemented in only specific and isolated segments of the business to "Digital Innovators," a state in which digital technology and thinking are integrated into every aspect of the organization.

When organizations are aware of where they stand on the Digital Maturity Model, they are better equipped to pinpoint areas of improvement and potential vulnerabilities. For instance, a company that finds itself only "Digital Initiator" might realize that it is missing out on harnessing the full potential of comprehensive digital integration, leading to lost opportunities or even competitive disadvantages. Conversely, those closer to the "Digital Innovator" end can validate their approaches and further optimize their strategies. Such clarity is essential not just for remedial action, but also for proactive strategizing.

By understanding their current standing on the Digital Maturity Model, businesses are empowered to craft tailored strategies that address their unique needs and challenges. This positioning serves as a foundational base, upon which companies can build their digital strategies, ensuring alignment with their current capabilities and future aspirations. Moreover, with this knowledge, organizations can set clear benchmarks, allowing them to track and measure their progress over time, adjusting their strategies as needed to ensure continued growth and adaptability in the ever-evolving digital landscape.

Infrastructure and data management overhaul

An overhaul of infrastructure and data management systems is imperative as part of a digital transformation roadmap for businesses aiming to stay relevant and competitive.

Central to this transformation is the shift toward cloud transition. Migrating data storage and operations to cloud platforms offers threefold benefits: enhanced accessibility allowing teams to access critical information anytime, anywhere; scalability, ensuring that the infrastructure can adapt to the growing demands of the business; and bolstered security measures, crucial for safeguarding sensitive data in an increasingly cyber-threat-filled environment.

Simultaneously, the incorporation of advanced data analytics tools is revolutionizing the way operations are approached and executed. These tools sift through vast amounts of data, extracting actionable insights that can significantly optimize project planning and outcomes. Such precision-driven decision-making is now the gold standard, ensuring that resources are utilized effectively and that businesses can anticipate and navigate challenges proactively. We shall delve into the detail of data-driven organizations in Chapter 5.

Complementing this is the adoption of unified communication and collaboration tools. Given the geographically dispersed nature of many teams, especially in large organizations, these tools are paramount. They ensure seamless collaboration across departments and locations, fostering an environment where information flows freely, and collective goals are achieved more efficiently.

Cultivating a digital-focused organizational culture

In the intricate process of digital transformation, organizational culture emerges as an often-underestimated factor. Many companies focus intently on technological investments and overlook the human element, not realizing that culture can either propel or hinder the evolution toward digital excellence. Organizations that are rooted in a culture resistant to change, or those that are overly cautious, frequently encounter roadblocks in their digital endeavors. For energy organizations, as well as others, the transformation journey must start with a culture overhaul, emphasizing curiosity, openness, and adaptability.

The drive toward a more digital-friendly culture necessitates changes at all levels of an organization. Leadership roles should evolve to become champions of change, advocating for innovation and demonstrating a willingness to adapt in the face of new technological realities. Simultaneously, there should be a concerted effort to invest in employee development, ensuring they are equipped with the skills and knowledge required in a digital age. Retraining and upskilling become essential components of this cultural shift, ensuring that the workforce is not only receptive to but also actively driving digital change. This involves the organization empowering its workforce to be digital citizens—to be digitally literate and can leverage digital tools and thinking in their everyday tasks. Training programs, workshops, and digital immersion experiences can cultivate a workforce that thinks, acts, and innovates digitally. This topic will be further explored in Chapter 8.

A truly transformative digital culture celebrates exploration and understands that growth often comes from trial and error. Organizations should prioritize the creation of an environment where experimentation is not just tolerated but actively encouraged. In such an atmosphere, failures are not seen as setbacks but rather as invaluable learning experiences. By adopting this mindset, companies can ensure that their teams are more resilient, innovative, and ready to tackle the challenges of the digital world head-on, turning potential stumbling blocks into stepping stones for future success.

Becoming data-driven

Modern energy organizations generate a staggering amount of data—from sensors on equipment, operations services performance data, to customer feedback. Much of these data is unstructured, meaning it does not fit neatly into traditional databases or spreadsheets. Harnessing these data requires sophisticated tools like machine learning algorithms and advanced analytics platforms. But beyond tools, organizations need to foster a data-centric mindset, where data-driven insights form the foundation of decision-making. This involves:

- Investing in infrastructure to store, process, and analyze data.
- Training employees in data science and analytics.
- Establishing protocols for data integrity, privacy, and security.
- Creating cross-functional teams that can derive actionable insights from data.

We shall explore what it takes to become data-driven in the next chapter.

Starting on your digital journey can be daunting. You are confronted, on the one hand, by the excitement of new, digitally enabled opportunities and, on the other, by the challenge of cultural inertia, the potential for misunderstanding between the IT function and the business, and uncertainty about whether your digital pilots will deliver.

By developing a robust digital transformation roadmap, energy organizations can navigate this transition more efficiently, unlocking new value streams, optimizing operations, and delivering superior stakeholder value. Embracing digital, understanding maturity, fostering the right culture, promoting digital citizenship, and harnessing data are the key pillars of this roadmap. With these in place, energy organizations can confidently stride into a digitally augmented future.

References

[1] Grand View Research, Grand View Research, by 2030, the Global AI Market Size Is Expected to Reach $1,811.8 Billion, Up from $136.6 Billion in 2022 with a 38.1% CAGR, 2023.
[2] J. McCarthy, John McCarthy, father of AI, Intell. Syst. IEEE 17 (5) (2020) 84–85.
[3] Offshore Technology: ADNOC Article, ADNOC Partners with G42 and Presight to Advance AI Solutions for Energy Sector, 2023.
[4] Shell Article, Shell Digitization Report, 2023.
[5] WEF—Digital Transformation of Industries, World Economic Forum (WEF)—Digital Transformation Initiative (DTI), 2022.

CHAPTER FIVE

Charting the future: The role of data-driven organization

Introduction

In an era where data are often heralded as the "new oil," a striking revelation emerges from the vast digital reserves: "Less than 1% of data in businesses is analyzed and turned into benefits," asserts global software giant SAP.

This startling statistic, set against a world that is increasingly digitized and data-reliant, underscores a significant disconnect between data and its actionable utilization. As industries, especially the transformative energy sector, navigate the complexities of the modern landscape, this underexplored data terrain holds the promise of untapped insights, innovative solutions, and strategic advantages. It beckons a pertinent question: Are we merely skimming the surface of the data-driven revolution?

The vast volumes of information generated from energy production to consumption hold the keys to reshaping energy futures, mitigating environmental impacts, and ensuring a resilient energy grid. Yet, the power is not just in the data, but in its intelligent application.

In the domain of unstructured data statistics, a noteworthy revelation is that approximately 90% of the data within the digital universe exists in an unstructured form. This indicates that the bulk of data generated does not conform to standardized formats conducive to easy searchability or sorting. In its initial, unprocessed state, these data present itself as disorderly and poses formidable challenges in terms of utilization. Nevertheless, it harbors significant intrinsic value.

Within this unstructured data, profound and actionable insights are often discovered, providing a rich source of information on aspects, such as consumer behavior, market trends, and operational inefficiencies. This highlights the utmost importance of developing sophisticated tools and

Powering Through the Transition
https://doi.org/10.1016/B978-0-323-91754-4.00004-2

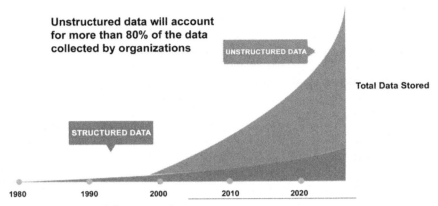

Fig. 5.1 Unstructured data—paradox.

technologies capable of effectively managing and deciphering this extensive, unregulated domain of unstructured data (Fig. 5.1).

Energy companies are not exempt from grappling with the unstructured data paradox. They also encounter the formidable task of handling and deriving value from the substantial volumes of unstructured data generated in their operations. As the energy sector continues its digital transformation and embraces data-driven approaches, the quantity of data collected from diverse sources, including sensors, satellite imagery, geological surveys, and maintenance reports, has grown exponentially. Yet, a significant portion of these data remains unstructured, presenting a challenge in fully exploiting its potential.

Unstructured data will constitute over 80% of the data amassed by companies.

This statistic highlights the reality that energy companies contend with a wealth of unstructured data encompassing unformatted text documents, sensor readings, images, and more. These data often harbor valuable insights regarding equipment performance, predictive maintenance, environmental factors, and market trends, all of which can profoundly impact operational efficiency and decision-making.

One of the primary obstacles energy companies face in harnessing unstructured data lies in the demand for advanced analytics and data processing tools. Traditional methods of data analysis prove ill-suited for handling the complexity and diversity inherent in unstructured data sources. To address this challenge, energy companies are increasingly turning to technologies, such as machine learning, artificial intelligence (AI), and big data analytics to extract actionable insights from unstructured data. These technologies facilitate predictive maintenance, real-time asset monitoring, and more informed decision-making, ultimately leading to improvements in operational efficiency and reductions in downtime.

The unstructured data paradox poses a significant challenge for energy companies as they grapple with the surging volume of unstructured data. Nevertheless, through the adoption of advanced analytics and data processing technologies, these companies can unlock the untapped potential of unstructured data, thereby enhancing their operations, reducing costs, and maintaining competitiveness in a continually evolving industry.

The process of extracting quality, ready-for-action insights from raw data takes a lot of time and effort. This has become the norm since, owing to much-advanced computing power and enhanced storage capacity, organizations are now faced with dealing with unprecedented amounts of unstructured or semistructured data, more than ever before. These data are generated from various sources like financial logs, text files, multimedia forms, sensors, and instruments. Simple data visualization tools are not capable of processing such huge volumes and variety of data. And so, organizations need more complex and advanced analytical tools and algorithms in order to process, analyze, and draw meaningful insights out of it.

This chapter takes a deep dive into the world of data-driven organizations within the energy sector. We will explore how these entities harness data to innovate, optimize, and lead in an era defined by both energy transition and digital revolution.

What does "data-driven" mean exactly?

As we venture deeper into the 21st century, the energy sector is confronted by unprecedented challenges and opportunities. Buffeted by environmental imperatives, shifting consumer demands, technological breakthroughs, and the monumental task of powering a rapidly increasing global populace, the traditional contours of the energy landscape are continuously evolving. Central to this evolution was an often-unseen influencer: data.

Today's transformative energy narrative is intrinsically interwoven with data. This vast reservoir of information, with its potential to streamline production, optimize operations, and forecast future energy trajectories, can be transformative. Yet, it is not merely about the data's accumulation. The true trailblazers are those entities that adeptly analyze, interpret, and act upon these data. These pioneering, data-driven organizations are the architects of an energy future that promises sustainability, efficiency, and adaptability.

Being a data-driven organization is more than just a buzzword or a modern mandate; it signifies a profound cultural shift. When a company treats data as a strategic asset, it fundamentally changes its perspective and prioritizes the sanctity, accuracy, and potential of every byte of information. This is not just about the colossal decisions that chart the trajectory of an enterprise but permeates down to the granular, everyday actions executed on the frontline. Employees, irrespective of their roles, become custodians of this asset, leveraging it in their daily tasks, challenges, and innovations. Such an organization does not just accumulate data; it breathes, lives, and thrives on it. It is about weaving data into the fabric of the company's DNA, ensuring that every decision, whether monumental or minute, is informed, optimized, and empowered by data-driven insights (Fig. 5.2).

In the dynamic landscape of the energy sector, where change is constant and technology is at the forefront, the term "data-driven" has emerged as a critical differentiator for businesses. In essence, a data-driven approach revolves around grounding decisions in concrete evidence amassed from data analysis.

At its core, a data-driven approach implies making decisions rooted in concrete, empirical evidence rather than intuition or tradition. It is about leveraging vast amounts of data to gain insights, predict trends, and optimize strategies. In the vast ecosystem of the energy sector, which spans from traditional oil and gas conglomerates to budding renewable energy enterprises, this methodology becomes crucial for several reasons.

For starters, the energy sector grapples with myriad challenges: complex and high-risk projects, fluctuating prices, intricate supply chains, stringent environmental regulations, and the growing demand for sustainable energy sources. As these companies endeavor to transition from fossil-fuel-based

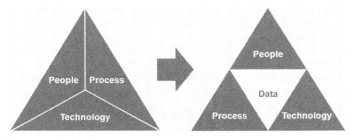

Fig. 5.2 Data as a strategic asset.

models to more renewable and sustainable ones, data provide a compass to navigate these turbulent waters.

In a data-driven company, decision-making at every level is predominantly influenced by data analysis and interpretation. This approach transforms how strategies are developed, operations are managed, and business decisions are made. Here are some key characteristics and implications of being a data-driven company:

Data collection

In data-driven companies, data collection is extensive and multifaceted, integrating information from a variety of sources. This can include real-time data from Internet of Things (IoT) devices, customer relationship management systems, and online analytics tools. It is not just about gathering large volumes of data but also about ensuring the diversity and relevance of the data collected. This includes both quantitative data (like sales figures, and project performance data) and qualitative insights (such as customer opinions or employee feedback). The emphasis is on creating a comprehensive data ecosystem that reflects various facets of the business and market.

Data analysis

The cornerstone of a data-driven company is its capacity to analyze complex data sets. This often involves sophisticated analytics tools, including AI and machine learning algorithms, which can identify patterns and insights that might not be apparent to human analysts. Organizations often establish dedicated teams or departments focused on data analysis, ensuring that these insights are not only generated but also effectively communicated across the organization. This integration of data analysis into all business functions means that every department, from marketing to finance, uses data to inform its strategies and operations.

Informed decision-making

Data-driven decision-making shapes both long-term strategies and immediate operational changes. By leveraging data, companies can identify growth opportunities, optimize their operations (such as refining supply chain processes), and proactively address potential risks. This approach enables a more strategic and less reactive business model, where decisions are based on evidence and analysis rather than intuition. The benefits extend beyond operational efficiencies to include enhanced innovation and competitive differentiation. Mining data for insights allows companies to identify underserved market segments or emerging customer preferences that would otherwise go unnoticed. In finance, for example, data-driven forecasting models enable more accurate financial planning and risk mitigation, ensuring that resources are allocated optimally. This evidence-based approach fosters greater confidence in decision-making at all organizational levels, from the C-suite to the front lines, allowing teams to execute strategies with clarity and conviction. In times of crisis, data-driven companies can pivot faster and more effectively, ensuring they stay ahead of disruptions.

Continuous learning

A key aspect of data-driven companies is their agility and capacity for continuous learning. They constantly refine their strategies based on new data and market trends, creating a dynamic business model that can quickly adapt to changes. This involves establishing feedback loops where data-driven actions inform future data collection and analysis, creating a cycle of continuous improvement. Continuous learning is powered by a company's ability to institutionalize learning from failures and successes alike. Tracking KPIs and outcomes enable organizations to identify what works and what doesn't, leading to iterative improvement. Moreover, the establishment of cross-functional teams, where data analysts work alongside business leaders, ensures that the learning process permeates the entire organization. This iterative approach turns experimentation into a structured, data-informed process where risks are minimized, and opportunities for improvement are consistently realized.

Cultural emphasis on data

In a data-driven company, data literacy is a core competency. Employees at all levels are trained to understand and use data effectively in their decision-making processes. Leadership plays a critical role in cultivating this culture, emphasizing the importance of data as a strategic asset and ensuring that its value is recognized and leveraged throughout the organization. To build this data-centric culture, organizations often invest in comprehensive training programs, ensuring that employees not only know how to use data tools but also understand the principles of data interpretation and analysis. Data democratization is crucial here, where access to relevant data is not restricted to a select few but is available across the company, empowering all levels of the organization to make data-backed decisions. Incentives and KPIs that reward data-driven actions further embed this culture. Moreover, fostering collaboration between technical teams, such as data scientists and business units helps bridge the gap between data insights and actionable outcomes, ensuring that data informs strategy at every level.

Ethical considerations

Ethical data practices are integral to the operations of data-driven companies. This includes ensuring the privacy and security of the data collected, especially sensitive customer information. Companies must navigate the complex landscape of data ethics, ensuring they comply with regulations (like GDPR) and maintain consumer trust by responsibly managing the data they collect and use. Beyond compliance, companies must also consider the broader ethical implications of how they use data. This involves avoiding bias in AI algorithms, ensuring that the data collected is representative and not skewed toward certain demographics. Transparency in how data is used is essential for maintaining trust; consumers increasingly demand to know how their data is being collected, stored, and utilized. Data-driven companies must also balance innovation with ethics, ensuring that new data strategies do not overstep societal or legal boundaries. Fostering an organizational culture of ethical responsibility enables these companies to mitigate the risk of data misuse as well as enhance their reputations as trustworthy stewards of information.

Why become data-driven?

By leveraging data as a strategic asset, a data-driven company gains a comprehensive understanding of its market dynamics, customer needs, and internal processes. This approach leads to objective, efficient, and innovative decision-making, setting the company apart in today's highly competitive and rapidly evolving business landscape. This transformation is not just about adopting new technologies; it is about a fundamental shift in the company's culture and approach to business.

So, what are the key advantages of becoming data-driven?

Firstly, the process of gathering and analyzing data significantly enhances confidence in decision-making across various business scenarios, such as decisions in project delivery or expanding into new markets. Data act as a vital benchmark against current operations, providing clarity on the potential impacts of various decisions. This is a stark contrast to reliance on intuition, as data offer a logical, tangible foundation for making decisions. This objectivity reduces subjectivity, thereby boosting confidence in choices made both at the individual level and across the organization. However, it is essential to acknowledge that decisions based on data are not foolproof. The accuracy of these decisions is contingent on precise data collection and correct interpretation. Therefore, continuous monitoring and evaluation of these decisions are imperative to maintain their effectiveness and relevance. We shall explore data-driven decision-making in the next section in a little more detail.

In the energy sector, data-driven strategies are particularly valuable for understanding and serving customers more effectively. Energy companies can analyze customer data to gain insights into consumption patterns, preferences for energy sources, and responsiveness to pricing changes. This information allows them to tailor their offerings, such as customized energy plans, demand-response programs, and renewable energy options, to meet the specific needs of different customer segments. For example, by identifying customers interested in sustainability, energy companies can offer green energy solutions or incentives for energy-efficient appliances. Personalized communication and services based on these insights can enhance customer satisfaction, leading to increased loyalty and retention. Additionally, data analysis can help energy providers discover new market opportunities, such as areas with a high demand for solar power installations or electric vehicle charging infrastructure. Continuously analyzing data enables energy companies to stay ahead of and act on industry trends. This proactive approach enables them to innovate and develop new products and services, ensuring they remain competitive and relevant in a rapidly evolving energy market.

Data analysis plays a crucial role in identifying and mitigating risks. The application of data analysis for risk management is increasingly more important and can be a major benefit for data-driven companies. Diligently examining trends, patterns, and outliers within their collected data drives energy companies to anticipate and address potential challenges before they evolve into significant issues. This forward-looking approach encompasses a wide range of business aspects, from managing financial and operational risks to mitigating reputational damage. For instance, by analyzing operational data, companies can foresee equipment malfunctions, minimizing the risk of power outages. Financial risks, like fluctuations in fuel prices, can be better managed through market data analysis. Furthermore, continuously monitoring environmental and safety data helps in predicting and preventing hazardous incidents, crucial in an industry where safety is paramount. Additionally, keeping a pulse on regulatory changes through data trends helps to ensure compliance and potentially avert legal or regulatory repercussions. By proactively identifying these risks, energy companies can develop and implement targeted strategies to either avoid them altogether or lessen their impact to ensure a more resilient and sustainable business.

The adoption of data-driven processes plays a pivotal role in enhancing the efficiency and responsiveness of a company's supply chain. This adoption can take many forms, from an end-to-end solution to adoption in targeted areas within a supply chain function. Systematically capturing supply chain data and analyzing it from different stages of the supply chain enables energy companies to identify critical bottlenecks that impede smooth operations. This insight allows for the implementation of targeted solutions to alleviate these issues and improve the profitability and efficiency of the function. Furthermore, accurate demand forecasting, which is made possible by analyzing historical consumption patterns and market trends, enables companies to manage their inventory levels more effectively. Optimizing inventory not only reduces operational costs but also ensures that resources are available when needed, thereby improving delivery times. Although this is a common principle, true data-driven companies may benefit significantly by collecting and utilizing data, which ultimately drives a reliable and efficient service. In essence, data-driven supply chain management is a key factor and advantage in achieving cost efficiency, operational excellence, and customer satisfaction.

Companies can also use data to enhance their sustainability efforts by analyzing data related to energy consumption, waste production, and resource utilization, which enables businesses to identify areas where they can reduce their environmental impact. This contributes to a more sustainable business model and aligns with increasing consumer and regulatory demands for environmentally responsible practices.

The benefits of data-driven decision-making extend far beyond improved confidence and cost savings. They encompass a wide range of business aspects, from customer engagement to innovation, risk management, human resources, supply chain optimization, and sustainability among many others. These advantages highlight the transformative power of data in shaping modern business strategies and operations.

The 2022 global survey conducted by NewVantage Partners was a significant endeavor aimed at examining the current landscape of data, AI, and data initiatives within major organizations across various industries. The survey's primary focus was to gauge the readiness and inclination of leading organizations to embrace a data-driven approach and develop a data-driven culture. The survey was especially relevant in 2022 as data-driven decision-making continued to gain prominence across the business world. Organizations were increasingly recognizing the transformative potential of data and AI, not only for improving operational efficiency but also for driving innovation, customer engagement, and competitive advantage. Hence, understanding the state of data initiatives and the commitment of organizations toward fostering a data-driven culture was crucial [1].

The survey highlighted some interesting themes among the participating organizations. Firstly, there was a widespread investment in data initiatives with over 90% of these organizations reportedly realizing measurable business benefits. This trend highlights the growing recognition of the tangible advantages that data and AI investments bring to businesses. It is noted that despite the investments, many organizations still struggle to fully embrace a data-driven culture with less than half of the participants indicated that they were effectively competing on data and analytics, and only a quarter have successfully created a data-driven organization. The statistics emphasize the clear ongoing challenges facing organizations that wish to adopt a data-driven approach and harness the full potential of data for strategic advantage.

The survey noted that AI initiatives were gaining momentum but still limited in scope. AI initiatives are on the rise, more than doubling and moving into widespread production (26%). The majority of these initiatives remain in pilot or limited production stages. This suggests that while interest and investment in AI are high, the full integration of AI into business operations is still an evolving process.

The cultural barrier to data-driven transformation was considered the main obstacle, with over 90% of executives highlighting cultural challenges as the main barrier in becoming data-driven. This is a key aspect of data-driven companies and can mean the difference between successful adoption and complete failure in adoption of data-driven strategies, which will be explored later in this chapter.

The survey considers the evolving role of a Chief Data and Analytics Officers (CDAOs). CDAO has become more prevalent in recent years, growing from 12% in 2012 to 73.7% in 2022 according to the survey. The CDAO typically holds a multifaceted role at the intersection of technology, data, and strategic leadership. CDAOs are entrusted with developing and executing an organization's data strategy, overseeing data governance, architecture, and analytics initiatives, and harnessing the power of data to drive business outcomes. They are responsible for cultivating a data-driven culture, ensuring data quality, and establishing robust data privacy and security measures. CDAOs may even use advanced analytics and machine learning to unearth valuable insights from vast datasets, enabling data-informed decision-making across all levels of the organization. They play a pivotal role in transforming raw data into actionable intelligence, driving innovation, and gaining competitive advantage by identifying emerging trends and opportunities. However, this role is still evolving, with many organizations reporting instability and turnover in these positions. The expansion of CDAO responsibilities to include business growth and analytics outcomes reflects the evolving nature of data leadership in organizations.

In the final aspect of the survey, a significant spotlight was cast on the area of data ethics, which is rapidly evolving into an indispensable focal point for organizations in the age of data and AI. The revelation that fewer than half of the surveyed organizations have well-established policies and practices governing data and AI ethics underscores the mounting awareness and urgency surrounding responsible data utilization and ethical considerations within the context of data and AI deployment. This shortfall in comprehensive ethical frameworks implies a heightened recognition of the ethical dilemmas, biases, and potential risks associated with data-driven technologies. As organizations strive to balance the immense benefits of data and AI with the ethical imperatives of fairness, transparency, and privacy, the findings signal a growing commitment to safeguarding not only data assets but also the rights and interests of individuals and communities affected by these technologies. It signifies a pivotal moment in which businesses are increasingly acknowledging the ethical dimensions of their data and AI endeavors and are poised to advance responsible and ethically sound practices in the rapidly evolving digital landscape.

The backbone of data-driven companies: Data-driven decision-making

Society has presented upon the notion of "intuition," the instinctive sense of right or wrong, which carries a high degree of prestige, importance, and influence. Recent studies have indicated that most professionals place their trust in their "gut feeling" when making decisions, even when confronted with conflicting evidence. The romanticization of intuition has permeated modern life to such an extent that it has become integral to how many people perceive and admire the intellectual giants of our era. In scientific circles, Albert Einstein's famous quote, "The intuitive mind is a sacred gift," is often quoted, while in the domain of business, Steve Jobs' quote to "have the courage to follow your heart and intuition" is frequently referenced.

While intuition can serve as a valuable initial motivation, it would be a mistake to base all decisions around a mere gut feeling in context of business decisions. While intuition can provide a stimulus that starts you down a particular path, it is through the lens of data that decisions can be substantiated, comprehended, and quantified. According to a survey of over 1000 senior executives conducted by PwC, organizations that prioritize data-driven decision-making are three times more likely to report substantial enhancements in their decision-making processes compared to those that place less emphasis on data [2].

Data-informed decision-making entails utilizing data to guide and validate decision-making processes before finalizing a course of action. In the business context, this manifests in various ways, such as:

Energy companies can use data to make informed decisions regarding the optimization of energy resources. For instance, by analyzing historical energy consumption patterns and weather data, companies can predict peak demand periods and allocate resources accordingly. This data-driven approach can help them optimize the deployment of power generation assets, ensuring they are running efficiently during high-demand periods and minimizing operational costs during low-demand times. Additionally, data analytics can assist in identifying opportunities for renewable energy integration, such as determining the most favorable locations for wind or solar farms based on historical weather and energy production data.

Another example that illustrates this point is in predictive maintenance, where data can play a crucial role. Collecting and analyzing real-time data from sensors placed on critical plant equipment such as turbines, transformers, and power lines allows energy companies to predict when maintenance is needed before equipment failure occurs. This proactive approach reduces downtime and costly emergency repairs and also enhances safety. Predictive maintenance can be guided by machine learning algorithms that consider historical performance data, environmental factors, and equipment conditions to forecast maintenance needs accurately. This data-driven strategy can result in substantial cost savings and operational efficiencies for the company.

The integration of data into the decision-making process hinges on several factors, including business objectives and the availability and quality of accessible data. While data collection and analysis have long been pivotal in larger corporations and organizations, the current era, with its daily generation of over 2.5 quintillion bytes of data, offers businesses of all sizes unprecedented opportunities to collect, analyze, and translate data into actionable insights. Although data-driven decision-making has been a part of business for centuries, it has truly come into its own in the modern age.

Becoming data-driven is more than just having access to a lot of data; it is about how that data are collected, analyzed, and integrated into every aspect of the organization's operations and culture. This approach leads to more informed decision-making, continuous adaptation and learning, and a commitment to ethical data practices. Data-driven decision-making stands as the cornerstone of success for companies within the energy sector. This approach involves a meticulous process of collecting, analyzing, and applying data to guide strategic and operational decisions. Understanding this process is crucial for any organization aspiring to harness the full potential of its data assets.

Mechanics of data-driven decision-making

The mechanics of data-driven decision-making form a comprehensive and structured process that transforms raw data into actionable business insights. This process begins with systematic data collection from a variety of sources. These sources range from direct customer interactions and market trends to internal operational metrics. The key here is to amass a diverse and comprehensive data set that reflects different facets of the business environment.

Once the data are collected, the crucial task of data management comes into play. Efficient data management involves ensuring the accuracy, relevance, and accessibility of the collected data. This step is critical because the quality of data directly impacts the reliability of the insights derived from it. Effective data management practices ensure that the data are not only current but also cleansed of any inaccuracies or inconsistencies.

The next phase involves advanced analytics, where the processed data are subjected to rigorous analysis using sophisticated tools and techniques. This stage often employs statistical analysis, predictive modeling, and machine learning algorithms to delve deep into the data. The goal here is to uncover hidden patterns, trends, and correlations that are not immediately apparent. This step transforms raw data into meaningful insights, revealing opportunities, challenges, and critical business drivers.

Insight generation is where the true value of data-driven decision-making comes to the forefront. The aim of this analysis is not just to crunch numbers but to generate actionable insights that can guide critical business decisions. It involves a deep understanding of the underlying patterns and trends within the data and translating these into practical business strategies.

The final step in this process is the implementation of these insights. This is where the rubber meets the road. The derived insights are applied to a range of decision-making scenarios, encompassing strategic planning, operational improvements, and tactical maneuvers. This step ensures that the insights do not just remain theoretical but are translated into concrete actions that can drive business growth and efficiency.

Data-driven decision-making is a meticulous process that turns data into a strategic asset. Systematically collecting, managing, analyzing, and applying data empowers organizations to make more informed, objective, and effective decisions, steering the company toward success in a competitive business landscape.

The balance between intuition and data

In the energy sector, characterized by its intricate and ever-changing dynamics, achieving a delicate equilibrium between intuition and data-driven decision-making stands as a paramount challenge. Data serve as a sturdy and objective foundation upon which to base critical decisions, ranging from forecasting energy demands to optimizing the allocation of valuable resources. Nevertheless, the intrinsic role of human intuition, refined through years of hands-on experience within the industry, should not be underestimated. The most effective decisions in the energy sector often emerge from the harmonious fusion of data-driven insights and the wisdom garnered from seasoned managerial expertise. This synergy ensures that decisions are not solely reliant on raw data but are also enriched by a profound comprehension of the distinctive challenges and opportunities that the energy sector presents.

The advantages of data-backed decisions in the energy sector are substantial. Data-driven decision-making minimizes the influence of personal biases and unfounded assumptions, yielding outcomes that are both objective and effective. Within an industry as pivotal and dynamic as energy, the capacity to employ data for predictive and prescriptive analytics holds immense value. These analytics empower energy companies to anticipate forthcoming market trends, predict energy demands, and assess resource availability, thereby enhancing strategic planning and resource management. Harnessing the power of data permits energy enterprises to become more adept at responding to prevailing market conditions and also gain the capability to strategically position themselves for future developments, ensuring sustainability and efficiency in their operations.

Nonetheless, the importance of human intuition in the decision-making process cannot be overstated. It serves as a vital complement to data-driven insights, allowing decision-makers to interpret and contextualize data within the unique framework of the energy sector.

The energy sector's decision-making landscape necessitates a finely tuned balance between data-driven insights and human intuition. While data offer precision and objectivity, intuition brings valuable industry knowledge and adaptability. The backbone of data-driven companies lies in their ability to effectively leverage data for informed decision-making, triking a balance between data and intuition, and embracing the power of data analytics to achieve greater efficiency, innovation, and competitiveness.

Transitioning to a proactive strategy in a data-driven environment is a critical evolution for businesses seeking a competitive edge. This shift involves moving from simply reacting to data insights into anticipating future trends and challenges. Here is a view of how companies can make this transition, along with examples:

Developing predictive analytics capabilities

Companies initially use data to understand current trends and past performance. However, by employing predictive analytics, they can forecast future trends and market dynamics. For example, project services companies can use data analytics to drive efficient project management practices, analyzing project timelines, resource utilization, and workforce efficiency to identify areas for improvement, reduce project delays, and optimize resource allocation.

Energy operators can use data analytics to predict when equipment or infrastructure might fail or require maintenance. For instance, a utility company could use sensor data from power lines and substations to predict and prevent outages, thereby reducing downtime and improving service reliability.

Implementing advanced data tools and technologies

The adoption of advanced data tools, such as AI and machine learning, can facilitate the transition to a proactive data management approach, by analyzing large data sets more efficiently and uncovering insights that might not be apparent through traditional analysis.

Operators of renewable energy sources, like wind farms or solar panels, can analyze weather data to predict energy production levels. This allows them to adjust operations and align with energy demand patterns, ensuring optimal energy production and reduced wastage. For instance, the UK is using AI and machine learning to help predict how much energy its turbines and solar panels will harvest when the wind blows or the sun shines brighter. Improved solar forecasts will improve the efficiency of the system and ultimately lower consumer bills.

Google also revealed in 2022 that thanks to AI it can predict the production of its wind farms in advance, allowing the technology giant to bid to supply power to the grid ahead of time and at lower values.

Integrating data insights into strategic planning ensures that decisions are forward-looking. Operating companies, for instance, might use data on global supply chain trends to inform its long-term sourcing strategy, thus avoiding potential disruptions.

Example: Utilizing data-driven decision-making

In the energy sector, data-driven decision-making is becoming increasingly important for large-scale energy corporations such as Shell. Shell's commitment to harnessing the power of data analytics exemplifies the transformative potential of this approach. By strategically deploying a comprehensive array of IoT sensors and data collection devices across its vast spectrum of operations encompassing exploration, production, and distribution, Shell is actively participating in the data-driven transformation within the energy sector.

These IoT sensors continuously capture an extensive range of critical data points, including wellhead pressure, pipeline flow rates, and the health status of various equipment components. This wealth of real-time data is then seamlessly transmitted to a centralized analytics platform where sophisticated algorithms come into play. These algorithms meticulously process and analyze the data, unveiling valuable insights that can drive more informed decision-making across the organization.

One notable application of data-driven decision-making at Shell revolves around predictive maintenance. Closely monitoring equipment performance data in real-time facilitates Shell to proactively predict when key assets like pumps, compressors, or drilling rigs may be vulnerable to potential failures. This forward-looking approach empowers the company to schedule maintenance activities with pinpoint accuracy, minimizing downtime, and substantially reducing maintenance costs. As a result, Shell's operations remain highly efficient, ensuring that critical equipment functions optimally.

Moreover, Shell's data analytics capabilities extend into the realm of optimizing energy production and distribution. For instance, the company leverages predictive analytics to forecast fluctuations in energy demand, drawing insights from historical data trends and weather patterns. Armed with this predictive capability, Shell can dynamically adapt its production and distribution strategies in real-time, thereby guaranteeing a nimble and responsive approach to meet energy demand efficiently, especially during periods of peak demand or unexpected market shifts.

How to create a data-driven culture

Cultivating a culture that values data-driven insights is crucial. Employees at all levels should be encouraged to think about how data can inform their work. Many organizations collect data and use it for limited monitoring efforts. But most organizations do not base any major business decisions on it. In a data-driven organization, data take a central role in the most important decisions. But even making selective use of data to inform decision-making can lead to rapid gains. For that reason, data-driven decision-making is not an all-or-nothing proposition. It does not require an overnight dramatic shift in the way you do business. That would be difficult to achieve even under the best of circumstances. Instead, you can gradually introduce a data-driven culture into your organization. Building it up over time leads to greater success. As measurable results start to emerge, the idea gains acceptance and even enthusiasm.

Creating a data-driven culture in your organization requires strong leadership. But it takes more than just a directive from the top. It needs an organization-wide commitment to make the transition. That begins with removing internal barriers.

Identify barriers within your organization

There can be various internal obstacles to adopting a data-driven approach within your organization. These barriers can manifest in different ways, ranging from outdated hardware to a deep-rooted reliance on intuition alone. Unfortunately, some stakeholders within the organization may have a mistrust of data. Consequently, there is a risk that they may hesitate to embrace a data-driven strategy.

Employees who are still using outdated tools and processes might also be resistant to change, as people tend to stick with familiar methods and procedures. Learning new skills can be challenging, and for change to occur, everyone involved needs to understand the benefits of a data-driven approach. It is crucial to effectively communicate these advantages throughout your organization.

Overcoming these cultural barriers may not be straightforward, but it is essential for your organization's transformation into a data-driven entity. Success requires a dedicated effort to address these challenges, but the rewards are substantial.

So, how can we address and mitigate these barriers?

Everything about a data-driven approach starts with collecting high-quality data. You must also make sure that it is accessible to everyone who needs to make a decision. If your organization is operating in silos, then that means data are not being freely shared. It is necessary to break down those silos and open up communication. All decision-makers across the organization need to access data.

A data-driven culture needs a single, unified source of accurate and trustworthy data. Create a central internal resource that explains how and where to find the data. That includes the content, format, and structure of your databases. This is essential to promote decision-making based on data. Urge everyone in your organization to ask questions. Make the resources available to give them the answers they need. Reward curiosity and make it a part of your culture.

Encouraging collaboration between departments can lead to a more holistic view of data and its implications. For instance, a collaboration between the sales and product development teams in a consumer electronics company might lead to the early identification of customer needs and the development of new products to meet those needs.

Training and development in data skills

Remember that no one is born knowing how to interpret data accurately. That is why training is such a vital part of any data-driven strategy. Keep in mind that training needs will vary throughout your organization. Front-line workers will need to be trained in different skill sets than management. Investing in employee training to enhance data literacy and analytical skills is key. This empowers employees to identify and act on opportunities pro-actively. For example, a healthcare provider could train its staff to use patient data to identify those at risk of chronic diseases and intervene early. Managers should have at least some basic training in statistics, for example. All training should build an understanding of how to use metrics, set goals, and arrive at conclusions correctly.

Data-driven decision-making often requires a fundamental change to the way people work. That often means investing in training materials and courses to impart the knowledge workers need. It may also mean developing new work structures over time. Everyone in your organization should know how to talk about data, visualize it, and present it when making decisions.

The final key is patience. The transition to a data-driven organization will not happen overnight. But you can nurture the process by using data to make small, tactical predictions in every area of your business. Take action based on these predictions, and then measure the results. Feed those results back into your process and use them to improve future decisions. Make these improvements visible throughout your organization. Focus not only on what happened, but why. The results will show the organization and executives the real value in a data-driven strategy, which in turn increases buy-in and helps organically spread the concept throughout your organization.

Evolution of a data-driven organization

The transformation of an organization from being data-resistant to achieving a state of data-driven is a profound and evolving journey that reflects changes in mindset, strategies, and organizational culture. It signifies the growing recognition that data are not merely a byproduct of business operations, but a powerful asset that can be leveraged to gain a competitive edge, make informed decisions, and drive innovation. There are numerous models that chart the evolution of a data-driven organization and provide insights into the challenges and opportunities organizations may encounter along the way.

Typically, the five stages of the evolution of a data-driven organization are illustrated in the figure next [3] (Fig. 5.3):

Stage 1: Data resistant

At the outset of this transformation journey, many organizations exhibit a level of skepticism and reluctance toward adopting a data-driven approach.

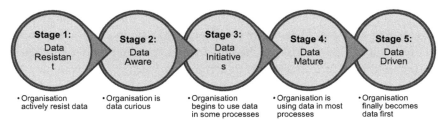

Fig. 5.3 Evolution of data-driven companies.

Decision-making predominantly relies on intuition and traditional method-ologies, with data playing a minimal role. The organization may lack the necessary infrastructure and expertise to work with data effectively. More-over, there might be a perception that data initiatives are costly and time-consuming, with uncertain returns on investment. Resistance to change is a common sentiment, as employees may feel threatened by the shift toward data-driven decision-making.

Overcoming the challenges of this stage involves building awareness about the potential benefits of data. It requires leadership to initiate conver-sations about the value of data-driven insights and to invest in the necessary resources, such as data infrastructure and training, to lay the foundation for a data-driven culture. Gaining buy-in from key stakeholders and demonstrat-ing the positive impact of data on decision-making can be pivotal in pro-gressing beyond this initial resistance.

Stage 2: Data aware

As the competitive landscape evolves and the advantages of data-driven decision-making become more apparent, organizations transition into the "Data Aware" stage. This shift is often triggered by external factors, such as market disruptions or witnessing the success of data-savvy competitors. Leadership begins to recognize the need for data, even if the organization is not yet fully equipped to harness its potential. Some initial investments may be made in data analytics tools and talent, albeit without a comprehen-sive strategy.

During this stage, organizations start to acknowledge that data can pro-vide valuable insights. They may experiment with data-driven projects in specific areas of their operations, such as marketing or inventory manage-ment. However, challenges may persist, including data silos, limited data lit-eracy, and a lack of standardized processes for collecting and analyzing data.

To advance further, organizations need to develop a clear data strategy that aligns with their business objectives. This includes defining KPIs and establishing a roadmap for data utilization. Education and training become crucial as employees need to acquire the skills necessary to work with data effectively.

Stage 3: Data initiatives

In the third stage of the transformation journey, organizations actively embrace data-driven initiatives. They invest in building the necessary data infrastructure, hire data professionals, and start collecting and analyzing data in a more structured manner. These initiatives are often project-specific and target areas, such as marketing optimization, supply chain efficiency, or customer insights. A noticeable shift occurs as data become recognized as an asset with the potential to drive tangible business improvements.

However, challenges persist as organizations scale their data initiatives. Data quality and governance issues may surface, requiring attention and investment. Organizations must also ensure that data-driven insights are integrated into decision-making processes, rather than existing as isolated projects. This involves fostering a culture of data-sharing and collaboration across teams and departments.

To progress beyond this stage, organizations need to elevate their data capabilities. This includes implementing advanced analytics tools and technologies to derive deeper insights from data. Additionally, they should focus on aligning data initiatives with broader business strategies and goals, ensuring that data-driven insights contribute to overall success.

Stage 4: Data mature

In the "Data Mature" stage, organizations have established a robust data culture and capabilities. Data-driven decision-making becomes pervasive across various functions and levels of the organization. The focus shifts from sporadic projects to more holistic data integration and utilization. Organizations prioritize data quality, security, and governance, recognizing their critical role in maintaining data integrity.

To excel in this stage, organizations invest in comprehensive data governance frameworks and data management practices. They develop standardized processes for data collection, storage, and analysis, ensuring consistency and reliability. Cross-functional collaboration becomes the norm, with data serving as a unifying language that informs decision-makers across the organization.

Continuous improvement is a hallmark of this stage. Organizations regularly assess their data strategies and adjust them to align with evolving business objectives. They harness data to identify market trends, customer preferences, and operational efficiencies. Success stories emerge as data-driven initiatives yield quantifiable benefits, such as increased revenue, reduced costs, and improved customer satisfaction.

Stage 5: Data-driven

The pinnacle of the data-driven evolution is the "Data-Driven" stage. Here, data are deeply ingrained in the organizational DNA, guiding every aspect of operations. The organization boasts a sophisticated analytics ecosystem, often incorporating advanced technologies like AI and machine learning. Continuous innovation is driven by data insights, and the organization leads its industry by leveraging data for breakthroughs in customer experience, product development, efficiency gains, and competitive advantage.

At this stage, organizations not only have advanced technical capabilities but also a culture that values data-driven decision-making as a strategic imperative. Employees at all levels are empowered with data access and insights, enabling them to contribute to the organization's success. Data-driven initiatives are not isolated but are integrated into the fabric of the organization, influencing every decision and action.

To attain and maintain this level of excellence, organizations must remain agile and adaptable. They need to continually embrace emerging technologies, foster a culture of innovation, and challenge the status quo. Data governance and security continue to be paramount concerns, as the volume and complexity of data grow. The organization serves as a benchmark for data-driven success, inspiring others in the industry to follow suit.

The evolution of a company from being data-resistant to achieving data-driven excellence is a dynamic and transformative journey that requires visionary leadership, strategic investments, and a commitment to cultural change. While the path may vary from one organization to another, the rewards of this transformation are substantial. Embracing data-driven decision-making can lead to sustainable growth, agility in responding to market changes, and superior decision-making in today's data-centric business landscape.

How to transform into a data-driven organization

In an era where data are becoming the lifeblood of business decision-making, the journey toward becoming a data-driven organization has emerged as a paramount strategic imperative.

Harnessing the power of data to inform and drive critical decisions, optimize processes, and unlock new opportunities is no longer an option but a necessity for long-term success. This transformative endeavor represents a profound shift in organizational culture, practices, and technologies.

This is a journey aimed at ensuring the alignment of analytics initiatives to organizational objectives, coupled with the seamless and efficient coordination of activities across all business units. The path from raw data to valuable insights, and subsequently from insights to actionable outcomes, is guided by strategic objectives. Frequently, organizations invest a significant portion of their time in the journey from raw data to insights, which constitutes a substantial part of their data-driven transformation efforts.

With the exponential growth of data access, it is evident that technological innovation will continue to disrupt traditional business models, industries, and global markets. In today's digitally driven global economy, the lines between the technology sector and other industries are rapidly blurring, if not entirely disappearing.

So how does an organization transform into a data-driven organization?

The journey starts with the why?

Just like any business scenario, organizations must invest time in comprehending the underlying reasons for their data needs. This fundamental query serves as the catalyst for formulating a comprehensive data-driven implementation strategy, ultimately ensuring a successful transformation. Moreover, it initiates a cascade of subsequent inquiries, each demanding meticulous consideration and resolution. Essentially, this exercise aims to maintain an organization's unwavering focus on grasping the overarching objectives and requisites from both a business and operational perspective.

Regrettably, many organizations often find themselves in a disconcerting situation. They rush to mobilize essential Business Intelligence resources without giving due regard to the data's quality, its alignment with the organization's culture, its role in the decision-making process, and its far-reaching impact on various departments. Ensuring the integrity of the data received is as paramount as "Garbage In, Garbage Out" holds true. Even the most exceptional model cannot yield superior results when fed with subpar data. It extends beyond the realm of hardware, software, infrastructure, and applications, delving into the critical realms of people and governance.

All too often, organizations fail to recognize the true value of the vast reservoirs of data at their disposal, let alone establish robust measures and controls for data quality and ethical handling. This underscores the imperative for organizations to center their attention on discerning the underlying purpose of data acquisition, as it forms the bedrock of their data governance policy.

In summary, for the majority of organizations, the impetus behind their data requirements arises from their aspiration to gain deeper insights into their customers or enhance their customer service and product offerings. It is from this core imperative that organizations can embark on the journey of implementing their data-driven strategy.

When formulating a data-driven strategy, organizations must adopt the perspective of treating data as a strategic asset and regard analytics as a critical strategic competency. While the term "analytics" encompasses a broad spectrum, it is paramount for organizations to discern the precise skillsets they require. To embark on a journey toward becoming data-driven, it is important to develop a strategy and roadmap from the outset. The following steps should be considered.

Assessment and planning

The first step on each organization's data-driven journey is to conduct a comprehensive assessment of its current state. This involves evaluating existing data practices, infrastructure, and capabilities. Understanding the present level of data maturity helps in identifying gaps and areas for improvement.

A key part when planning for your organization's data-driven journey is setting specific, measurable objectives for guiding the data-driven transformation. These objectives may include enhancing operational efficiency, reducing environmental impact, or improving safety measures. Clear goals ensure that the transformation is aligned with the company's overall strategic vision. Next, a detailed roadmap is needed, which lays out the path to achieving these objectives. This plan should include timelines, milestones, and KPIs to track progress. It should also outline the resources required, including personnel, technology, and budgets.

Building the technical foundation

Investing in modern data collection technologies, such as IoT sensors and advanced data analytics platforms, is a fundamental part of any data-driven strategy. This infrastructure serves as the foundation for gathering, storing, and processing vast amounts of operational data and may include the adoption of advanced analytics tools, for big data analytics, AI, and machine learning. This in turn enables the data-driven organizations to derive actionable insights from their data. These tools can help in predictive maintenance, optimization of operations, and risk management. The data infrastructure needs a detailed assessment based on the organizations data-driven strategy to ensure it adds value to the business and that consideration is given to ensuring integration and interoperability of the existing IT systems, which is key to a cohesive and efficient data-driven environment.

Cultural and organizational change

A successful transformation requires a shift in the organizational culture to value data as a strategic asset. This involves promoting a mindset that recognizes the importance of data in decision-making at all levels of the organization. In turn, training programs are needed to enhance employees' skills in data analytics and digital technologies, ensuring the workforce is prepared to engage with the new data-driven processes. Encouraging collaboration between different departments also facilitates the effective use of data. Shared data practices can lead to more holistic insights and innovation across the organization.

Leadership and governance

For a successful transition into a data-driven organization, it is imperative that the leadership team is fully committed to and actively supports data-centric strategies and initiatives. The role of leaders extends beyond mere approval; they need to be the driving force and advocates for these data-driven approaches, establishing a culture that values and utilizes data throughout the organization. Their engagement sets a tone that resonates across all levels, signaling the importance and urgency of this shift toward a data-centric model.

Equally important is the establishment of robust data governance policies. These policies play a critical role in maintaining data quality, ensuring data security, and adhering to relevant regulations and standards. They provide a structured framework for how data are collected, stored, accessed, and used within the organization, underlining the importance of responsible data management.

Furthermore, allocating sufficient resources and budget to data initiatives is a tangible demonstration of the organization's commitment to becoming data-driven. This investment is crucial for acquiring the necessary tools, technologies, and talent to effectively leverage data and drive organizational transformation.

Implementing data-driven solutions

The journey toward becoming a data-driven organization often begins with the implementation of small-scale pilot projects. These initial ventures serve as a test bed for data initiatives, allowing the organization to explore and learn from these efforts in a controlled, low-risk environment. Starting small is important for organizations to identify best practices, understand potential challenges, and gauge the effectiveness of data-driven solutions. When these pilot projects prove successful, they set a precedent and provide a scalable model that can be applied more extensively across the organization, thereby laying the groundwork for a broader transformation.

Following the success of these pilot projects, the organization should focus on methodically scaling and integrating data-driven practices throughout its various departments and functions. This expansion needs to be strategically planned to ensure that the integration of data-driven solutions is seamless and aligns with the overall business objectives.

Establishing a framework for ongoing monitoring and evaluation is important, utilizing KPIs to measure the impact and effectiveness of these initiatives. Regular assessment allows for the identification of areas for improvement, ensuring that the data-driven strategies remain aligned with the evolving needs and goals of the organization.

Continuous improvement and innovation

To successfully transform into a data-driven organization, it is important to adopt an iterative approach centered around continuous improvement. This philosophy entails regularly refining and enhancing data practices, ensuring they evolve in response to new insights, feedback, and results. Constantly seeking ways to improve empowers organizations to ensure that its data-driven strategies remain effective, relevant, and aligned with its evolving objectives. This approach fosters a dynamic environment where learning from outcomes and adapting strategies is a continual process, vital for the long-term success of data-driven initiatives.

Keeping up to date with the latest advancements in data analytics and technology is another key aspect of this transformation. Staying informed about emerging trends and innovative practices in the data realm enables organizations to continuously enhance its data capabilities and maintain a competitive edge. Moreover, cultivating a workplace culture that encourages and rewards innovative uses of data is essential. This environment of creativity and experimentation allows for the exploration of novel ideas and solutions, driving the organization toward cutting-edge applications of data. Such a culture not only inspires new ways of thinking but also reinforces the organization's commitment to being at the forefront of data-driven innovation.

Addressing challenges and resistance

Transforming into a data-driven organization involves fundamental changes in organizational processes and mindsets, which can naturally be met with resistance. To effectively navigate this transition, it is important to have a robust change management strategy in place. This should focus on understanding and addressing the concerns of employees, providing adequate training, and supporting the organization through the transition. Change management is not just about introducing new technologies or processes, and it is about guiding the organization through the change, helping them understand the value and implications of these new data-driven practices. Successful change management ensures that the transition is not only smooth but also sustainable, with widespread acceptance and adherence to new methodologies.

Another critical aspect of this transformation is maintaining clear and transparent communication across the organization. Open dialogue about the goals, processes, and expected benefits of becoming a data-driven organization is vital for securing widespread buy-in. Employees at all levels should be made aware of how this shift will impact their roles and the overall success of the company. Demystifying the transformation process and its objectives is important for organizations to mitigate uncertainties and foster a positive attitude toward change. Clear communication also involves listening to feedback and addressing any concerns, ensuring that the transition to data-driven practices is inclusive and considers the perspectives and needs of the entire workforce.

A significant challenge in the shift toward being a data-driven organization is the skill gap that might exist among employees. To address this, it is crucial to invest in training and development programs. These programs should be designed to equip employees with the necessary skills and knowledge to effectively engage with new data tools and practices. Providing comprehensive training ensures that all team members are not only comfortable with but also proficient in using data analytics and interpreting data-driven insights. This investment in human capital not only helps in overcoming resistance by empowering employees but also enhances the overall efficiency and effectiveness of the organization's data-driven initiatives. By fostering a culture of continuous learning, the organization can ensure that its workforce remains agile and capable of adapting to evolving data technologies and methodologies.

An additional challenge in the transition to a data-driven organization is breaking down existing silos within the company. Data often reside in isolated departments, leading to segmented insights and a lack of cohesive strategy. To address this, fostering a culture of collaboration across different departments is essential. Implementing cross-functional teams and platforms where data can be shared and analyzed collectively can help in achieving a more integrated approach. This collaborative environment encourages different parts of the organization to work together, leveraging data to drive unified decision-making and strategy. It is key to break down these silos so that organizations can optimize the use of its data as well as cultivate a more inclusive and synergistic work culture, essential for the successful adoption of data-driven practices.

Partnering and collaborating

One of the strategies that should be considered in the journey toward becoming a data-driven organization is to form strategic partnerships with external experts, such as technology providers, institutions, consultants, and industry specialists. These alliances can bring in a wealth of specialized domain knowledge and skills that may not be present internally within the organization. Leveraging this external expertise enables organizations to gain access to advanced technologies, innovative methodologies, and industry-specific insights. These partnerships can be particularly valuable in areas where the organization is looking to enhance its capabilities or accelerate its data-driven initiatives. They not only fill knowledge gaps but also provide fresh perspectives that can drive more informed decision-making and strategic planning.

In addition to forming external partnerships, it is equally important for the organization to actively engage with a broader ecosystem of stakeholders, including suppliers, customers, and industry groups. Such collaboration can lead to a richer, more diverse pool of data and insights. Interacting and sharing data with these stakeholders allow organizations to gain a more comprehensive understanding of market trends, customer needs, and supply chain dynamics. This broader engagement not only enhances the quality and scope of data-driven strategies but also opens up new opportunities for cooperation and innovation in the industry. For energy companies, embracing this collaborative approach is crucial in leveraging data to optimize operations, drive innovation, and remain competitive in the ever-evolving energy sector.

Getting the insights

For energy companies, the journey toward becoming data-driven has advanced beyond the stages of data collection and understanding. The real challenge now lies in extracting meaningful insights from data. Despite having well-organized data systems, insights do not flow automatically, especially when dealing with nontraditional data sources for example, IoT sensors deployed on equipment in the field. It is therefore important for these organizations to focus on cultivating the right data skills. As highlighted in recent industry reports, many organizations are facing a shortfall in meeting their data skills requirements, emphasizing the need for a workforce proficient in transforming data into actionable insights. Transforming into a data-driven culture means that the reality of data drives every major work effort. When everyone from the executive team down makes data-driven decisions, then the entire organization benefits. The investments made in a data-driven approach pay immense dividends in the future. Once your data-driven organization is on the right path, you are set to reap the rewards of data-driven decisions. The growing agility of your organization and increasing effectiveness of your decisions are well worth the effort.

Adopting a data-driven strategy does not necessitate a complete departure from intuition or subjective opinions. Instead, it involves balancing data interpretation with common sense and experience to glean data insights. The concept of a data-driven approach is not reserved for only the biggest companies in the world, nor is it an all-or-nothing approach. Today, organizations of any size can benefit from the transformative power of data and analytics and it does not mean that you have to abandon intuition or ignore opinions. In fact, the ability to interpret data using common sense and experience is arguably as important as the data itself.

Developing a data-driven strategy is a patient and challenging task but critical for future success. Implementing an organization-wide data-driven transformation, which encompasses strategy, organizational structure, analytics, and technology, can be a daunting task for both management and employees. Historical trends suggest that more than half of such comprehensive transformation initiatives fail to meet their economic and timeline objectives. This transformation requires a dedicated commitment, often taking months or years to realize tangible benefits, and may face significant resistance from within the organization. Leadership is key in driving a data-driven strategy forward, a topic to be expanded in Chapter 10. It is crucial to have dedicated leaders at both the enterprise and departmental levels to steer this transformation.

References

[1] NewVantage Partners Data and AI Executive Survey, NewVantage Partners 2022 Data and AI Executive Survey, 2022.
[2] 10Minutes, Balance the Art of Instinct with the Science of Data and Analysis, 2015, May.
[3] C.S. Penn, The Evolution of the Data-Driven Company, 2005, July. https://www.christopherspenn.com/2019/08/the-evolution-of-the-data-driven-company/.

CHAPTER SIX

Center of Excellence

Introduction

As the consequences of climate change become increasingly evident, nations, industries, and individuals are seeking ways to reduce their carbon footprint and transition to cleaner and more sustainable sources of energy. Paradoxically, for energy companies, this transition represents a monumental challenge and a unique opportunity. To navigate this complex and dynamic landscape successfully, they must embrace innovation, data-driven decision-making, and digitization.

This transition is fraught with challenges, including the need for major investments in infrastructure, the complexity of integrating intermittent renewables into the energy grid, the changing consumer preferences for cleaner energy sources, and all this while embracing the new era of digital technology, big data, and artificial intelligence (AI). In the face of these challenges, energy companies need a strategic approach that can harness their existing expertise while facilitating the adoption of these new technologies and practices. This is where Centers of Excellence (CoEs) come into play. CoEs are specialized units within organizations dedicated to fostering excellence in specific domains or functions. While their roles and scopes can vary widely, their core mission is to drive innovation, standardization, and knowledge sharing.

CoEs serve as hubs of expertise, bringing together cross-functional teams with specialized knowledge and skills. In the context of energy companies navigating the energy transition, CoEs are poised to make a major impact. They act as catalysts for change by creating a culture of continuous learning and improvement, facilitating the adaptation of cutting-edge technologies, and fostering collaboration across the organization.

The concept of CoE in many ways mirrors the principle of "sharpening the saw" as advocated by Stephen Covey in his influential book, "The 7 Habits of Highly Effective People." Covey's metaphor for self-renewal, self-care, and continuous improvement is a powerful lens through which

Powering Through the Transition
https://doi.org/10.1016/B978-0-323-91754-4.00002-9

to view the establishment and operation of a CoE, especially in the context of the rapidly evolving energy sector. Just as Covey emphasizes the importance of preserving and enhancing our greatest asset, oneself, through balanced self-renewal, a CoE serves as a focal point for an energy company to continuously refine and expand its capabilities, technologies, and methodologies. This is particularly pertinent as the industry navigates the complexities of the energy transition, moving toward more sustainable and renewable sources of energy. The CoE acts as a hub for innovation, knowledge sharing, and skills development, ensuring that the organization remains at the forefront of industry advancements and is well-equipped to adapt to changing market demands and environmental considerations. In the face of the energy transition, the analogy extends to the CoE's role in fostering resilience and adaptability within the organization. Just as "sharpening the saw" involves a proactive approach to self-improvement across physical, social/emotional, mental, and spiritual dimensions, a CoE adopts a holistic approach to organizational development. It drives the pursuit of excellence through research, development, and the implementation of best practices in energy production, efficiency, and sustainability. By doing so, the CoE not only enhances the company's ability to innovate and compete but also aligns its operations with the broader goals of the energy transition. This alignment is crucial for energy companies looking to thrive in a future where the emphasis on renewable resources, carbon footprint reduction, and environmental stewardship is ever-increasing.

In this chapter, we will delve into the theory and practical application of CoEs in energy companies. We will begin by establishing the theoretical foundation of CoEs, explaining their historical evolution, and highlighting the importance of tailoring CoEs to suit the unique needs of energy organizations. We will then explore the synergies between CoEs and digitization, data-driven approaches, and innovation. As we embark on this journey, it is evident that CoEs are not just organizational units; they are a strategic imperative for energy companies seeking to thrive in the energy transition. Their potential to drive innovation, harness data-driven insights, and tailor solutions to the unique needs of the energy sector position them as essential tools for the future of energy. In the following sections, we will explore these themes in greater depth, shedding light on the transformative power of CoEs in energy companies navigating the energy transition.

Center of everything...

At its core, a CoE is a centralized and specialized unit within an organization that excels in a particular domain or function. The term "excellence" signifies a commitment to achieving the highest standards of performance, innovation, and knowledge dissemination in that specific area. CoEs are typically staffed by experts with deep knowledge and experience, and their mission is to drive excellence across the organization by promoting best practices, standardization, and continuous improvement. This group hones expertise in a specific subject area, standardizes best practices for wide-scale adoption, and provides thought leadership and direction in their area of expertise.

According to Gartner, CoEs exist to "concentrate existing expertise and resources in a discipline or capability to attain and sustain world-class performance and value." These virtual or physical centers combine learning and oversight in a specific area, driving the organization to shift across multiple disciplines together [1].

CoEs have become pivotal in spearheading innovations and efficiencies in the energy transition and can focus on a wide range of areas, such as renewable energy technology, grid optimization, energy efficiency, carbon reduction strategies, or even digital transformation. Each CoE is tailored to address the unique challenges and opportunities within its designated domain.

For instance, a CoE dedicated to renewable energy sources within an energy company could centralize expertise and knowledge in solar and wind power generation. This hub of knowledge works to refine the technology and operational strategies for harnessing these energy sources and to disseminate best practices across the entire organization to effectively raise the bar. Through a CoE, the organization can more effectively integrate renewable energy into its portfolio, navigating the complexities of variable power generation and storage solutions.

A CoE focused on grid modernization within an energy company plays a crucial role in transitioning the traditional energy grid into a smart grid. By centralizing expertise in Internet of Things (IoT) technologies, data analytics, and cybersecurity, the CoE can guide the organization through the intricate process of upgrading its grid infrastructure. This could include the deployment of smart meters, the integration of distributed energy resources, and the implementation of advanced predictive maintenance techniques. Through these efforts, the CoE supports the organization on its journey to enhance the reliability and efficiency of energy supply.

When deployed effectively, CoEs have the potential to make a monumental difference within an organization, serving as pivotal drivers for innovation, efficiency, and competitive advantage. Centralizing expertise, knowledge, and best practices in a specific domain enables CoEs to significantly enhance an organization's capabilities, streamline processes, and foster a culture of continuous improvement and excellence. However, the success of CoEs hinges on careful management, and a well-defined and focused remit. They must maintain a sharp focus on its core domain to be truly effective and deliver on its promise. The essence of a CoE lies in its ability to concentrate specialized skills, expertise, and resources to drive innovation, efficiency, and leadership in a specific field. Without this focus, there is a risk that it may dilute its impact by attempting to cover too broad an array of topics or functions, essentially becoming a "center of everything." This lack of specificity can lead to scattered resources, unclear objectives, and ultimately, a failure to provide tangible value in any one area. It is crucial for a CoE to define its scope with precision, ensuring it neither strays too far from its expertise nor stretches its capabilities too thinly. In doing so, it can avoid the pitfalls of overgeneralization and remain a beacon of excellence and a catalyst for meaningful advancements within its designated area.

An enabler to drive superior performance

CoEs are, in essence, catalysts for organizational excellence and are generally set up to drive superior organizational performance, often through innovation. They serve as dynamic hubs that drive performance improvement, foster innovation, and ensure that an organization remains adaptable in a rapidly changing environment, such as the energy transition. Here, we will delve into how CoEs embody key theoretical concepts and principles that underpin their effectiveness.

CoEs are built on the foundational principle of knowledge management, which emphasizes the importance of capturing, sharing, and applying knowledge within an organization. They are knowledge hubs, embodying principles of knowledge creation, sharing, and dissemination. They leverage the organization's collective intelligence to drive innovation and problem-solving. Knowledge management theories emphasize the importance of capturing and applying organizational knowledge effectively.

In the context of the energy sector, this means that CoEs become repositories of specialized knowledge. They house subject matter experts who possess deep insights into the challenges and opportunities specific to the energy transition. These experts curate and disseminate knowledge to other parts of the organization, ensuring that best practices are adopted consistently.

CoEs play a pivotal role in fostering a culture of innovation within an organization. Innovation management theories highlight the need for dedicated structures like CoEs to promote idea generation, experimentation, and the development of new solutions. CoEs are often seen as innovation incubators, driving the adoption of cutting-edge technologies and practices. Innovation is central to the energy transition, as it requires the development and adoption of new technologies, processes, and business models. CoEs provide the ideal environment for innovation to thrive. They encourage out-of-the-box thinking, experimentation, and collaboration across departments. Through innovation, CoEs help energy companies discover novel solutions to complex problems.

CoEs play an important role in leading organizational transformation. They serve as the catalysts for introducing groundbreaking ideas, state-of-the-art technologies, and innovative practices that challenge the status quo. The success of such transformative initiatives hinges on effective change management strategies. These strategies are essential for fostering acceptance of new changes, rather than encountering resistance. CoEs leverage these change management techniques to clearly articulate the advantages of adopting new changes, actively engage with various stakeholders, and cultivate a culture of ownership and commitment throughout the organization. This holistic approach ensures that the innovations introduced are not only accepted but also fully integrated into the organizational fabric. Acting as change agents, CoEs embody the principles of change management to navigate the organization through the complexities of adopting novel practices and technologies. Their role involves a strategic blend of managing resistance to change, effectively communicating the positive impacts of change, and facilitating a seamless transition process. By doing so, they ensure that the transition is not just a mere adoption of new practices but a transformative journey that enhances the organization's overall efficiency and competitiveness. Through their focused efforts, CoEs play a crucial role in guiding the organization through the intricacies of change, ensuring that the journey is marked by shared understanding, enthusiastic participation, and a collective stride toward a more innovative and agile organizational structure.

The philosophy of continuous improvement is fundamentally ingrained in the operations of CoEs, shaping their approach to achieving excellence in every aspect of their work. CoEs constantly seek opportunities to enhance operational efficiency, minimize environmental impacts, and achieve cost optimization. This mindset drives them to meticulously identify areas for improvement, strategically implement necessary changes, and rigorously measure the outcomes of such initiatives. Through this cycle of perpetual refinement, CoEs enable energy companies to maintain their competitive edge and adapt fluidly to the dynamic shifts in market demands and regulatory landscapes. Embracing continuous improvement if important for CoEs to optimize their current operations and anticipate and prepare for future challenges.

Furthermore, CoEs play a crucial role in fostering a culture of continuous improvement within organizations. This culture is built on the foundation of recognizing the value of incremental progress, optimizing processes for greater efficiency, and leveraging insights gained from previous experiences to inform future strategies. The principles of continuous improvement championed by CoEs encourage a proactive stance toward problem-solving and innovation. CoEs can provide a collaborative platform to facilitate the exchange of ideas, analysis of performance data, and the collective refinement of methodologies. This environment nurtures a mindset among team members that is geared toward constant learning and adaptation, ensuring that the pursuit of excellence is a shared organizational endeavor.

The impact of instilling a continuous improvement ethos through CoEs extends beyond operational enhancements; it transforms how organizations view challenges and opportunities. If deployed correctly CoEs can empower teams to approach tasks with a critical eye, always searching for ways to do things better, smarter, and more sustainably. This approach leads to a virtuous cycle of innovation, where small, incremental improvements accumulate over time to yield significant advancements. Such a culture does not merely adjust to changes in the external environment but actively shapes the organization to be more resilient, agile, and forward-thinking. In essence, the drive for continuous improvement cultivated by CoEs is a key ingredient in building organizations that are not just equipped to survive but thrive in an ever-evolving global marketplace.

The CoE has become instrumental in driving value across business units within organizations, thanks to its specialized subject matter expertise and operational agility. CoEs excel in harnessing and disseminating capabilities, fostering innovation, and elevating overall organizational performance. Below are five major avenues through which CoEs deliver value.

Innovative thought leader

CoEs add substantial value to organizations by acting as innovative thought leaders. Within their structure, CoEs have the unique role of pioneering new ideas and approaches in their specialty areas, thereby setting the pace for innovation across the organization. Fostering a culture that is keen on exploration and creative problem-solving enables CoEs to encourage continuous innovation and adaptation. This leadership in innovation ensures that the organization keeps up with industry trends and stays ahead of them maintaining a competitive edge. CoEs serve as a beacon for innovative practices, attracting and nurturing talent that is predisposed to forward-thinking and disruptive technology. This environment of intellectual stimulation and cutting-edge research propels the entire organization toward groundbreaking solutions and services.

As innovative thought leaders, CoEs contribute by transforming cutting-edge ideas into practical applications that enhance business operations and customer satisfaction. They often lead pilot projects that test and refine new concepts before wider implementation across the organization. This role includes introducing novel solutions and assessing their impact and effectiveness, which informs strategic decisions and investments. Bridging the gap between theoretical advancements and practical implementations can help CoEs to ensure innovations translate into real-world benefits, driving sustained organizational growth and performance improvements. Their ongoing commitment to innovation helps to instill a corporate ethos that values and seeks out continual improvement and adaptability, preparing the entire organization to meet future challenges with resilience and agility.

Centralization and optimization of resources

CoEs play a pivotal role in optimizing resources within organizations, centralizing scarce and highly sought-after assets like specialized knowledge, skills, and experiences. This strategy of centralization not only broadens the accessibility of these crucial resources across the organization but also streamlines the process of leveraging them effectively. By acting as a central hub for critical assets, CoEs ensure that every department and business unit can efficiently tap into a pooled repository of expertise, skills, and knowledge. This eliminates redundant efforts and siloed operations, fostering a more unified and cohesive approach to organizational challenges. The result is a more streamlined resource allocation model that promotes efficiency and reduces waste, allowing projects and initiatives to proceed with enhanced operational effectiveness.

Moreover, CoEs significantly enhance collaboration and knowledge sharing across the organization. By bringing together experts from various disciplines, CoEs create a dynamic environment where ideas and strategies can be exchanged freely and innovation can flourish. This cross-pollination of knowledge not only sparks new ideas and solutions but also ensures that these innovations are grounded in a deep understanding of diverse organizational needs and perspectives. The collaborative nature of CoEs breaks down traditional barriers between departments, promoting a culture of openness and collective problem-solving that is essential for agile and adaptive business practices. In turn, this environment supports the development of a more knowledgeable and versatile workforce, equipped to handle the complexities of today's business landscape and drive the organization forward with a unified vision and purpose.

Faster delivery

In the rapidly evolving energy sector, the ability to expedite the delivery of projects and initiatives significantly boosts a company's competitive edge. CoEs play a crucial role in this aspect by eliminating the common bottlenecks that typically hinder the swift access to vital capabilities and resources. CoEs can create a streamlined pathway to these critical resources, which can facilitate faster and more effective decision-making. The result is a marked improvement in the organization's responsiveness to market fluctuations and opportunities ensuring that products and services are developed and improved at a pace that aligns with industry demands and maintained to meet ongoing changes and challenges. This faster delivery mechanism extends beyond mere speed, enhancing the organization's agility and ability to innovate in response to market needs.

By providing quick access to specialized knowledge and technological advancements, CoEs empower teams across the organization to prototype, test, and refine solutions in shorter cycles, which in turn leads to more rapid product and services iterations and enhancements, as well as ensures that these innovations are more closely aligned with customer expectations and emerging market trends. Ultimately, CoEs help solidify the organization's position at the forefront of the industry, driving growth through agility and strategic responsiveness.

Cost optimization

One of the most tangible benefits of CoEs is their ability to drive cost efficiencies. Through the elimination of inefficient practices, CoEs streamline processes, foster the creation of reusable assets, and reduce redundancy across the organization. This approach to operational efficiency not only decreases direct costs but also optimizes resource utilization, leading to significant savings. The focus on creating reusable assets and methodologies means that solutions developed in one part of the organization can be adapted and applied elsewhere, reducing the need for redundant efforts and enabling more strategic allocation of financial and human resources.

Quality of services and products

CoEs play a critical role in standardizing best practices across the organization, which in turn ensures a consistent and high-quality delivery of services and products. This standardization process involves the establishment of benchmarks, guidelines, and quality control measures that are applied uniformly, leading to improvements in the reliability and quality of outputs. Additionally, by fostering a culture of excellence and continuous improvement, CoEs contribute to tightening end-to-end customer experiences. This focus on quality not only enhances customer satisfaction but also bolsters the organization's reputation in the marketplace, driving long-term success and sustainability.

CoEs serve as a linchpin for organizational excellence, offering countless benefits that extend from operational efficiencies to enhanced quality and customer satisfaction. Centralizing expertise, streamlining processes, optimizing costs, and standardizing best practices empower CoEs to support organizations to navigate the complexities of the energy transition with agility and confidence.

Case study: Saudi Aramco

A great illustration of how energy companies have embraced the concept of CoEs and used them as a vehicle to drive innovation, resulting in differentiated service offerings is Saudi Aramco's Exploration and Petroleum Engineering Center of Excellence (EXPEC), which stands as an exemplary model of how CoEs can serve as critical differentiators in the energy sector. Established in the early 21st century, EXPEC highlights Saudi Aramco's commitment to pushing the boundaries of oil and gas exploration and production. Centralizing research, development, and innovative technologies has enabled EXPEC to play a pivotal role in enhancing the company's operational efficiency and sustainability. Its focus on leveraging cutting-edge technologies and methodologies for exploration and production operations has not only streamlined processes but also significantly reduced

environmental impact, setting new standards in the industry. The establish-
ment of EXPEC aligns with the broader strategic objectives of Saudi
Aramco to maintain its leadership position in the global energy market. This
CoE concentrates on developing proprietary technologies and methodologies that can unlock new reserves, improve recovery rates, and optimize the
production lifecycle. In doing so, EXPEC has effectively propelled Saudi
Aramco to the forefront of technological advancements in the energy sector.
Its success demonstrates the power of a well-focused CoE in driving innovation that directly contributes to the company's bottom line and resilience
in the face of fluctuating market conditions and environmental
challenges [2].

A specific example that illustrates the impact of a CoE in the energy sector, particularly within Saudi Aramco, is its pioneering work in advanced
seismic technology through EXPEC. Saudi Aramco's EXPEC Advanced
Research Center (EXPEC ARC) has developed cutting-edge seismic technologies that have revolutionized the way the company explores and evaluates its oil and gas reservoirs. One notable innovation is the development of
the 4D Seismic Monitoring technology, which enables the dynamic obser-
vation of changes in reservoirs over time. This technology allows Saudi
Aramco to predict fluid movements and reservoir conditions more accurately, leading to more efficient extraction strategies and enhanced recovery
rates. The deployment of 4D Seismic Monitoring technology by EXPEC
ARC exemplifies how a focused CoE can significantly advance operational
capabilities and industry standards. Prior to its development, the ability to
monitor reservoirs dynamically was limited, often leading to less optimized
extraction, and increased environmental impact. The introduction of this
technology not only improved operational efficiency but also reduced the
environmental footprint by minimizing unnecessary drilling and optimizing
resource extraction. This breakthrough has set new benchmarks in
the industry for sustainable resource management and operational
excellence [2].

EXPEC's role extends beyond the immediate operational improvements; it embodies Saudi Aramco's strategic investment in human capital
and knowledge sharing. Fostering a culture of continuous learning and innovation enables EXPEC to attract top talent and nurtures expertise within the
company. This emphasis on knowledge and skill development ensures that
Saudi Aramco remains adaptable and forward-thinking, capable of navigating the complexities of the global energy landscape. In this way, EXPEC not
only differentiates Saudi Aramco within the highly competitive energy

sector but also establishes a blueprint for how CoEs can be instrumental in driving sustainable growth and innovation in energy companies worldwide [3].

The versatility and adaptability of CoEs in addressing evolving challenges and opportunities within the energy sector is an enabler to drive to superior performance. CoEs need to continuously evolve to remain at the forefront of innovation and excellence.

Evolution of CoEs

CoEs have a rich and multifaceted history, tracing their origins back to the mid-20th century. The emergence and proliferation of CoEs across various sectors can be attributed to a convergence of influential factors that shaped their evolution.

In the aftermath of World War II, the world observed the extensive damage inflicted on economies and infrastructures on a global scale. There was a critical need for recovery and revitalization, leading to a period of remarkable rebuilding initiatives worldwide. Recognizing the monumental task at hand, governments and military organizations took decisive steps to establish CoEs as instrumental pillars in the postwar recovery process. These CoEs emerged as vital hubs for research, development, and specialized training initiatives, serving as focal points for expertise and innovation in countless fields essential to the reconstruction endeavor. Drawing upon a wealth of intellectual capital and resources, these institutions spearheaded pioneering efforts aimed at revitalizing economies, restoring infrastructure, and fostering social and economic stability in the wake of the war's devastation.

CoEs similarly played a major role for military institutions in propelling military capabilities forward, honing strategic tactics, and fortifying operational readiness in the postwar period. These centers became linchpins in the revitalization and streamlining of armed forces. Their contributions proved instrumental in modernizing military forces, guaranteeing adaptability to dynamic security landscapes and geopolitical pressures.

During the mid-20th century due to the rapid pace of scientific and technological advancements, CoEs found their place as specialized units capable of effectively managing and disseminating knowledge in emerging fields, such as aerospace, nuclear energy, and information technology. CoEs swiftly emerged as crucial mechanisms for harnessing expertise in these cutting-edge domains, thereby facilitating innovation and technological progress. By fostering collaboration between academia, industry, and government agencies, these centers facilitated the rapid development and deployment of breakthrough technologies, spurring economic growth and enhancing national competitiveness on the global stage.

Furthermore, the advent of the quality management movement, championed by renowned figures, such as W. Edwards Deming and Joseph M. Juran, exerted a profound influence on the evolution of CoEs. The pioneering principles and methodologies advocated by these pioneers played a pivotal role in reshaping the organizational landscape, prompting the establishment of specialized CoEs dedicated to enhancing quality and standardization across industries [4]. W. Edwards Deming, often hailed as the father of quality management, emphasized the importance of statistical process control, continuous improvement, and the cultivation of a culture of quality within organizations and similarly, Joseph M. Juran advocated for the concept of quality management as a strategic imperative for achieving organizational excellence, emphasizing the need for systematic approaches to quality improvement and the engagement of all stakeholders in the process. Drawing upon the insights and methodologies espoused by Deming, Juran, and other quality management pioneers, organizations across various industries began to recognize the value of establishing CoEs focused on quality enhancement. These centers served as focal points for disseminating best practices, providing training and education, and driving initiatives aimed at improving quality standards and processes.

As CoEs began to establish their foothold primarily in sectors like defense and manufacturing, their significance quickly transcended these traditional domains and expanded into the energy sector. This transition was not merely coincidental but stemmed from the inherent suitability of CoEs to address the intricate challenges inherent to the energy sector. The energy sector is characterized by its complexity, volatility, and rapid evolution, driven by factors, such as technological innovation, regulatory changes, and shifts in consumer demand. Among these dynamics, the adoption of CoEs emerged as a strategic imperative for energy companies seeking to navigate the complexities of the industry and stay ahead of the curve. One significant factor contributing to the relevance of CoEs in the energy sector is the growing emphasis on innovation and technological advancement.

The energy sector's additionally increasing focus on sustainability and environmental stewardship further emphasized the importance of CoEs. As organizations grappled with the challenges of climate change, resource depletion, and regulatory pressures, CoEs specializing in sustainability initiatives and green technologies emerged as critical drivers of change. These centers played a pivotal role in advancing sustainable practices, driving the adoption of renewable energy solutions, and spearheading initiatives to reduce carbon emissions and minimize environmental impact. The advent of renewable energy marked a significant milestone for CoEs within the energy sector. As global interest shifted toward cleaner and more sustainable energy sources, specialized CoEs focusing on renewables emerged to drive the development, integration, and optimization of clean energy technologies. CoEs provided the ideal platform for fostering innovation, facilitating research and development initiatives, and driving the integration of new technologies into energy systems.

The dawn of the digital transformation era heralded a profound shift in the energy sector, ushering in a new wave of CoEs dedicated to digitization, data analytics, and cybersecurity. These specialized CoEs emerged as vital entities within energy companies, tasked with harnessing the power of digital technologies to revolutionize operations and fortify cybersecurity defenses in an increasingly digitized landscape. These CoEs spearheaded initiatives to modernize legacy systems and streamline processes, and enhance operational efficiency through the adoption of digital tools and technologies. By leveraging data analytics, these CoEs enabled energy companies to extract valuable insights from vast datasets, driving informed decision-making, optimizing resource allocation, and identifying opportunities for innovation and growth. Furthermore, cybersecurity CoEs played a pivotal role in safeguarding energy infrastructure and data assets against evolving cyber threats. With the proliferation of interconnected systems and the rise of cyberattacks targeting critical infrastructure, these CoEs implemented robust cybersecurity frameworks, conducted risk assessments, and deployed advanced defense mechanisms to protect against breaches, intrusions, and data breaches. CoEs dedicated to data analytics empowered energy companies to capitalize on the vast amounts of data generated across their operations. Employing advanced analytics techniques, such as predictive modeling, machine learning, and AI, enables these CoEs to unlock actionable insights, proactive maintenance, predictive analytics, and optimization of asset performance.

CoEs in driving the energy transition

As energy companies find themselves in the midst of navigating the intricate challenges of the energy transition, CoEs arise as indispensable strategic assets. This period of transition demands significant change in established business paradigms, technological landscapes, and operational methodologies. CoEs can provide a structured and systematic approach to navigating this transformation effectively, serving as the cornerstone between conventional energy operations and the pressing imperatives of a sustainable, environmentally responsible energy ecosystem.

CoEs are instrumental in helping energy companies to successfully navigate the complexities of the energy transition by providing strategic direction, fostering innovation, and facilitating collaboration. As organizations strive to meet the challenges of a rapidly evolving energy landscape, CoEs serve as catalysts for change, driving the adoption of sustainable practices and paving the way for a more resilient and environmentally responsible energy future.

Strategic enablers in the energy transition

In the broader context of the energy transition, CoEs emerge as strategic enablers, playing a pivotal role in empowering energy companies capitalize on the numerous opportunities presented by this transformative period. These CoEs fulfill specific roles that are essential for driving progress and facilitating successful transitions within the energy sector, some of the key roles are presented next.

- Facilitate cross-functional collaboration

  Facilitating cross-functional collaboration lies at the heart of CoEs within the energy sector. CoEs can create dynamic environment primed for collaborative innovation and problem-solving by assembling a diverse array of experts spanning various disciplines as needed by each organization in question, including engineering, environmental science, data analytics, and finance. These interdisciplinary hubs enable professionals with diverse backgrounds to converge and tackle the multifaceted challenges inherent in the energy transition. Engineers bring technical expertise in designing and implementing sustainable energy solutions, while environmental scientists may contribute insights into the ecological impacts of energy operations. Data analytics specialists harness the power of big data to derive actionable insights, inform decision-making, and optimize energy processes. Meanwhile, finance experts lend their financial acumen to ensure the viability and sustainability of energy projects, etc.

  This cross-disciplinary approach enables CoEs to address the complex and interconnected nature of energy challenges comprehensively, promoting dialogue, collaboration, and knowledge sharing. This enables

CoEs to foster a holistic understanding of the energy transition and facilitate the development of innovative, integrated solutions.

Furthermore, the collaborative environment nurtured by CoEs encourages creative thinking, out-of-the-box problem-solving, and the exploration of unconventional ideas. These diverse perspectives and expertise of its members result in a breakdown of organizational silos to overcome organizational barriers, and drive meaningful progress toward a sustainable energy future.

Overall, the ability of CoEs to facilitate cross-functional collaboration is instrumental in accelerating the pace of innovation, fostering synergy among stakeholders, and ultimately, driving positive change within the energy sector. As energy companies navigate the complexities of the energy transition, CoEs serve as catalysts for collaborative action, driving collective efforts toward a more sustainable and resilient energy ecosystem.

- Drive innovation

CoEs serve as dynamic engines of innovation, providing dedicated spaces where innovative thinking is nurtured, and novel ideas are explored. CoEs create environments conducive to driving transformative innovation across various facets of energy operations by fostering a culture of creativity, experimentation, and risk-taking.

One key aspect of CoEs' role in driving innovation is their dedication to exploring new technologies. These centers serve as incubators for emerging technologies, such as renewable energy systems, energy storage solutions, smart grid technologies, and advanced analytics tools. CoEs are set up to support companies to stay abreast of technological advancements and create an environment to foster innovation and experimentation, this enable companies leverage cutting-edge technologies and enhance efficiency, reliability, and sustainability.

Moreover, CoEs play a pivotal role in exploring innovative business models that can revolutionize the energy sector. From decentralized energy systems and peer-to-peer energy trading platforms to subscription-based energy services and energy-as-a-service models, CoEs actively explore and evaluate new business paradigms that can disrupt traditional energy markets and unlock new value streams.

In addition to technological and business model innovation, CoEs are at the forefront of driving sustainability initiatives within the energy sector. These centers spearhead research and development efforts aimed at advancing sustainable energy solutions, reducing carbon emissions, and mitigating environmental impact. From exploring novel approaches to renewable energy generation and storage to developing innovative strategies for energy efficiency and conservation, CoEs play a pivotal role in advancing the sustainability agenda within energy companies.

Furthermore, CoEs serve as platforms for collaboration and knowledge sharing, facilitating partnerships with external stakeholders, such as academia, research institutions, startups, and industry associations. Fostering open innovation ecosystems allows CoEs to drive companies to tap into external expertise, leverage diverse perspectives, and accelerate the pace of innovation. CoEs can also drive innovation by providing dedicated spaces for exploring new technologies, business models, and sustainability initiatives. Fostering a culture of creativity, experimentation, and collaboration enables CoEs to support their organizations in catalyzing transformative change, adapt to evolving market dynamics, meet sustainability goals, and maintain a competitive edge in an increasingly complex and dynamic landscape.

- Leverage data and technology
CoEs can act as instrumental platforms to leverage the power of data and technology to drive innovation and optimization of business processes and technology. These CoEs play a critical role in harnessing data-driven insights and digital technologies to optimize operations, enhance efficiency, and promote sustainability across the energy value chain.

At the heart of CoEs' efforts in this regard lies the utilization of advanced data analytics techniques to extract actionable insights from vast amounts of data generated by energy operations. Employing machine learning algorithms, predictive analytics models, and data visualization tools CoEs can analyze complex datasets to identify patterns, trends, and anomalies that can inform decision-making and drive operational improvements.

CoEs can also leverage digital technologies to optimize energy production processes. Through the implementation of smart sensors, IoT devices, and automation systems, CoEs enable real-time monitoring and control of energy infrastructure, allowing for proactive maintenance, predictive analytics, and optimization of asset performance. This proactive approach helps energy companies minimize downtime, reduce maintenance costs, and maximize the efficiency of their operations.

- Standardize best practices
CoEs are instrumental in establishing and standardizing best practices within energy companies to drive consistency, efficiency, and effectiveness in the adoption of sustainable energy solutions across the

organization. Developing standardized practices and frameworks enables CoEs to help guide companies in their journey toward sustainability and resilience. One key aspect of CoEs' role in standardizing best practices is the development of frameworks that outline guidelines, procedures, and methodologies for implementing sustainable energy solutions, for example. These frameworks may serve as reference documents, providing clear and structured guidance on how to design, deploy, and manage sustainable energy projects. Companies can streamline processes, minimize errors, and optimize resource allocation, ultimately driving efficiency and effectiveness in their operations and sustainability initiatives.

CoEs also play an important role in ensuring consistency in the adoption of sustainable energy solutions across different departments and business units within the organization. CoEs can support by promoting a unified approach to sustainability facilitating alignment and coordination among various stakeholders, ensuring that everyone is working toward common goals and objectives. This consistency helps to eliminate silos, foster collaboration, and maximize the impact of sustainability efforts across the organization.

CoEs facilitate knowledge sharing and capacity building initiatives to empower employees with the skills and competencies needed to implement best practices effectively. Through training programs, workshops, and knowledge transfer sessions, CoEs equip employees with the tools, techniques, and knowledge required to excel in their roles and contribute to the organization's sustainability agenda.

CoE customization

While CoEs provide a robust framework for attaining excellence for organizations navigating the uncertain journey of the energy transition, it is important to recognize that a universal approach will not suffice, particularly considering the intricate nature of the energy sector. Each organization faces its distinct challenges, opportunities, and operational contexts, necessitating customized CoEs tailored to their specific needs and circumstances. This customization ensures that CoEs are finely tuned to address the unique requirements of each organization to their effectiveness in driving organizational success amidst the ongoing transition.

Energy companies, whether operating in oil and gas, renewables, or utilities, function within unique ecosystems shaped by diverse regulations, market dynamics, and technological landscapes. As such, organizations that decide to implement a CoE cannot follow a cookie-cutter approach.

Customization is paramount because the energy sector is multifaceted, with challenges and opportunities that vary widely across different activities, including exploration, production, distribution, and a wide range of services and suppliers. Each of these activities presents its unique set of challenges, regulatory requirements, and technological demands.

For instance, traditional oil and gas companies may need to grapple with issues related to reservoir management, drilling efficiency, and emissions reduction. Traditional oil and gas companies are also navigating the energy transition by diversifying their portfolios and reducing their carbon footprint. CoEs in this context may focus on technologies like carbon capture and storage, methane emissions reduction, and the development of clean hydrogen production processes to bridge the gap between existing hydrocarbon operations and emerging sustainable technologies.

Conversely, a renewable energy company might prioritize optimizing the performance of solar farms, wind turbines, or energy storage systems. Customization is particularly critical in the renewable energy subsector, given the diverse range of renewable energy sources, such as solar, wind, and geothermal. A CoE focused on wind energy, for example, may specialize in optimizing wind turbine design, maintenance strategies, and grid integration for wind farms. Similarly, a solar energy CoE might concentrate on enhancing photovoltaic panel efficiency, reducing production costs, and developing innovative energy storage solutions specific to solar power.

In addition, it goes without saying that the energy transition is not a uniform process with different regions and countries at varying stages of adopting clean energy technologies and policies. CoEs must adapt to these regional differences, regulatory frameworks, and market conditions to remain effective.

Alignment with organizational objectives

CoEs should closely align with an organization's specific goals and objectives to ensure they contribute directly to the strategic vision that they were established to support. For energy companies striving to lead in innovation, establishing CoEs with a strong focus on innovation can be game-changing. These CoEs essentially become innovation hubs, fostering a culture of creativity and exploration within the organization. Employees are encouraged to push boundaries, experiment with emerging technologies, and collaborate with external partners, such as startups and research institutions. By staying at the forefront of technological advancements, these companies can continuously drive innovation in the energy sector, striving to position themselves as industry leaders.

For example, organizations that are committed to sustainability recognize the importance of integrating environmental considerations into their operations. CoEs dedicated to sustainability initiatives play a crucial role in this regard by focusing on strategies for carbon reduction, promoting circular economy practices, and enhancing environmental sustainability reporting. Aligning closely with the organization's sustainability goals enables CoEs to help drive meaningful progress toward becoming environmentally responsible energy providers, whereas organizations looking to improve operational efficiency and cost-effectiveness can benefit significantly from establishing CoEs dedicated to efficiency optimization. These CoEs can employ various methodologies, such as lean principles, business process reengineering, and data-driven analytics to identify inefficiencies in the organization and streamline operations. Optimizing the organizations workflows and processes can reduce costs and improve overall operational performance, enabling companies to be more productive.

In essence, customized CoEs that closely align with organizational goals and objectives serve as strategic enablers, driving innovation, sustainability, and operational excellence within energy companies. Tailoring CoEs to specific focus areas enables organizations to leverage their strengths and resources more effectively, ultimately driving long-term success.

Leadership and governance

Customized CoEs require robust leadership and governance structures to ensure alignment with the organization's strategic objectives. Leaders within CoEs who possess deep expertise in the relevant domains should be appointed, have a clear vision of how the CoE can contribute to the organization's objectives, and are capable of guiding teams effectively. They should have well-defined objectives that align with the organization's strategic plan, ensuring that the CoE's efforts are directed toward tangible outcomes.

Governance structures should oversee the CoE's operations, monitor progress, and ensure accountability, with boards or committees providing guidance, reviewing performance metrics, and making strategic decisions. CoE leaders should also continuously assess and ensure that their initiatives are aligned with the organization's evolving objectives, being agile and capable of adjusting the CoE's focus as organizational goals shift.

Collaborative ecosystems

In addition to internal customization efforts, energy organizations must recognize the immense value in fostering collaborative ecosystems that extend beyond their immediate boundaries. By forging strategic partnerships with external entities, such as research institutions, industry associations, and startups, organizations can significantly enhance the capabilities and the impact of their CoEs. This proactive approach to collaboration enriches the knowledge base and resources available to CoEs and also accelerates their contributions to organizational objectives.

Collaborating with universities and research institutions represents a cornerstone of this ecosystem-building strategy. Tapping into the wealth of cutting-edge research, academic expertise, and innovative solutions housed within these institutions enables CoEs to gain access to a reservoir of knowledge and capabilities that can accelerate their initiatives. Through collaborative projects and joint research endeavors, organizations can leverage academic insights to drive technology development, solve complex industry challenges, and stay at the forefront of innovation. Furthermore, partnerships with research institutions facilitate ongoing knowledge sharing and skill transfer, empowering CoEs to continuously evolve and adapt in response to evolving industry dynamics.

Membership in industry associations serves as another opportunity for collaboration and knowledge exchange. Actively participating in industry networks and consortia supports CoEs to gain valuable access to networking opportunities, industry best practices, and the latest trends and developments shaping the energy sector. Engaging with peers and industry experts through these associations facilitates cross-pollination of ideas, fosters collaborative problem-solving, and promotes the sharing of lessons learned and success stories. Furthermore, participation in industry associations provides CoEs with a platform to contribute to the development of industry standards and regulations, thus playing a proactive role in driving industry-wide improvements and advancements.

Additionally, integrating startups into the collaborative ecosystem can inject fresh perspectives, agility, and innovation into CoE initiatives. Startups often possess niche expertise, disruptive technologies, and an inclination for experimentation that can complement the strengths of established energy organizations. Partnering with startups through incubator programs, accelerator initiatives, or strategic collaborations allows CoEs to catalyze the development and adoption of innovative solutions, accelerate time-to-market for new products and services, and foster a culture of entrepreneurialism within the organization.

Building a robust collaborative ecosystem around CoEs empowers energy organizations to leverage external expertise, resources, and innovation ecosystems to amplify the impact of their initiatives. Harnessing the collective strength of academia, industry associations, and startups enables CoEs to drive transformative change, drive industry leadership, and achieve sustainable growth.

Flexibility and adaptability

Recognizing the dynamic and evolving nature of the energy transition, CoEs must be designed with inherent flexibility and adaptability. This proactive approach enables CoEs to navigate the complexities of the transition effectively and capitalize on emerging opportunities. Implementing agile methodologies lies at the core of this strategy, facilitating rapid responses to evolving circumstances and fostering a culture of innovation and resilience within CoEs. This entails embracing iterative development processes, where CoEs break down complex initiatives into smaller, manageable tasks or iterations. This approach allows for incremental progress, enabling CoEs to deliver value to stakeholders more frequently and adapt their strategies based on real-time feedback. Regular stakeholder engagement and feedback loops further enhance agility, ensuring that CoEs remain aligned with evolving stakeholder needs and expectations. Additionally, adaptive planning practices enable CoEs to adjust their priorities and resource allocations dynamically in response to changing market conditions, technological advancements, or regulatory requirements.

Engaging in scenario planning exercises represents another critical component of CoE adaptability. Exploring various hypothetical future scenarios and their potential implications permits CoEs to proactively identify risks, opportunities, and strategic pathways. Scenario planning drives CoEs to develop contingency plans, anticipate market shifts, and position themselves strategically to capitalize on emerging trends. This forward-thinking approach empowers CoEs to remain agile and resilient in the face of uncertainty and mitigate risks or maximize opportunities for success.

Fostering a culture of continuous learning and adaptation is important to sustaining CoE effectiveness. Prioritizing professional development initiatives, knowledge-sharing platforms, and cross-functional collaboration make it possible for CoEs to cultivate a learning-centric environment where team members stay up to date on industry trends, technological advancements, and regulatory changes. This ongoing investment in skills development and knowledge acquisition equips CoE personnel with the tools and insights needed to innovate, problem-solve, and adapt to evolving industry dynamics. Embracing a growth mindset encourages team members to explore new ideas, challenge conventional thinking, and drive continuous improvement across all facets of their operations.

Customization is an important aspect to consider when implementing a CoE, which needs a great deal of thought to be effective ultimately allowing them to align with organizational objectives, and remain agile and adaptable. This requires strong leadership, robust governance structures, collaborative ecosystems, and a focus on flexibility and continuous learning.

Measuring CoE success

As we have seen, CoEs stand as pivotal hubs for innovation, expertise, and strategic advancement. These specialized entities play a fundamental role in shaping the trajectory of energy companies from exploration, production, project services, sustainability, and many others. However, the effectiveness of CoEs in driving organizational success hinges upon their ability to perform optimally and adapt to dynamic industry landscapes. Hence, the measurement of CoE performance is important to ensure it is consistently adding value. Systematic evaluation of the effectiveness of CoE initiatives, processes, and outcomes is important so that organizations can gauge their impact, identify areas for enhancement, and ensure alignment with strategic objectives. In addition, recognizing that excellence is not a static achievement but a continuous journey, the ongoing monitoring and improvement of CoE performance become essential practices.

Measuring CoE performance poses unique challenges. Unlike tangible production metrics or financial indicators, the evaluation of CoE effectiveness often involves intangible aspects, such as knowledge dissemination, innovation diffusion, and organizational culture enhancement. Consequently, establishing KPIs requires careful consideration to ensure they encapsulate the multifaceted contributions of CoEs accurately. KPIs need to be meticulously developed to reflect not only short-term outputs but also long-term impacts on efficiency, sustainability, and strategic positioning. During the process of identifying and developing effective performance measures for CoE performance, ensuring that the right things are being measured needs to consider more than just easily quantifiable metrics, and is it important to consider qualitative aspects that drive CoE value as well.

It is prudent to develop a robust measurement framework to ensure that CoEs continue to perform optimally and deliver tangible value to the organization. Aligning KPIs with strategic objectives allows energy companies to assess the contribution of their CoEs to overarching goals, such as operational excellence, innovation leadership, and environmental stewardship etc.

Fostering a culture of continuous improvement also encourages CoEs to adapt, innovate, and evolve in response to evolving industry dynamics, technological advancements, and stakeholder expectations. Through this holistic approach to performance measurement and enhancement, organizations can get the most out of their CoEs.

Continuous improvement within CoEs

In the pursuit of operational excellence and sustained value delivery, CoEs must engage in a dynamic process of continuous improvement. This process encompasses several key concepts aimed at enhancing efficiency, efficacy, and alignment with organizational objectives.

Firstly periodic audits, which serve as foundational pillars of a continuous improvement cycle within CoEs. These audits entail comprehensive assessments of CoE activities and outputs to ensure operational effectiveness and efficiency. Through rigorous examination, audits evaluate compliance with established processes, adherence to industry standards, and the attainment of performance targets. The audits also serve as quality control mechanisms, to validate CoE operations, as well as unearth areas that need improvement, facilitating targeted interventions to bolster overall performance.

Establishing robust feedback loops with CoE stakeholders is important for informed performance assessment and refinement of CoE activities. This entails engaging employees, operations teams, and end-users of CoE-driven solutions to gain invaluable insights into the impact of CoE initiatives. Feedback mechanisms include regular surveys, focus groups sessions, or direct communication channels and enable stakeholders to articulate their experiences, challenges, and suggestions. This rich feedback informs decision-making processes within CoEs, enabling them to make timely adjustments and course corrections to better meet stakeholder needs and expectations.

Embracing an iterative approach to goal setting is also important for CoEs to remain agile and responsive. As organizational priorities evolve and industry challenges unfold, CoEs must adapt their objectives and initiatives accordingly and by continually reassessing and refining their goals, CoEs ensure alignment with overarching organizational strategies while proactively addressing emerging opportunities and threats. This iterative goal-setting process tends toward a culture of adaptability and innovation within CoEs, empowering them to drive meaningful change and value creation.

Continuous improvement within CoEs rises above the optimization of an organization's processes, it embodies a rounded commitment to excellence, innovation, and stakeholder satisfaction. Embracing regular audits, feedback loops, and iterative goal setting supports CoEs to work through complex problems, seize opportunities, and drive toward sustained success in the rapidly evolving energy industry landscape.

Demonstrating return on investment (ROI)

Quantifying the ROI and demonstrating the value of CoEs are important for organizations seeking to justify continued investment and earn support for their initiatives. Quantitative ROI calculations serve as a foundational pillar in assessing the financial effectiveness of CoE initiatives. These calculations involve a systematic comparison of the tangible financial gains and cost savings derived from CoE activities against the resources invested in their implementation.

For instance, a CoE focused on renewable energy might conduct a quantitative ROI analysis to evaluate the impact of its initiatives, such as the development of more efficient wind turbines. Quantifying metrics may be defined as reduced manufacturing costs, increased energy production, or enhanced operational efficiency through the implementation of CoE initiatives. In addition to quantitative ROI, organizations must also recognize and emphasize the qualitative value that CoEs contribute. Qualitative value encompasses a spectrum of intangible benefits that may not be immediately quantifiable in monetary terms but are nonetheless crucial indicators of CoE effectiveness and impact. These qualitative benefits include enhancements in organizational culture, heightened employee engagement, bolstered brand reputation, and meaningful contributions to sustainability and social responsibility goals. For instance, a CoE dedicated to sustainability initiatives may not yield immediate financial returns, but it can showcase the company's commitment to environmental stewardship and enhancing its brand image in the eyes of stakeholders.

Qualitative value may well extend beyond internal organizational dynamics to encompass broader societal and environmental impacts. CoEs that champion sustainability and innovation initiatives can play instrumental roles in driving industry-wide transformations, shaping regulatory frameworks, and advancing toward a more sustainable energy future. CoEs can position themselves as drivers of positive change, earning recognition and support from stakeholders by aligning their efforts with overarching societal needs and environmental imperatives.

Performance measurement tools and metrics for CoEs

An accurate assessment of the impact that CoEs can have on organizations requires a careful selection of performance measurement tools and metrics. There are numerous tools and metrics available, as highlighted next, and each plays a distinct role in capturing various dimensions of CoE performance providing valuable insights for continuous improvement efforts.

The Balanced Scorecard is a strategic planning and management system that can be used to align business activities to the vision and strategy of the organization, improve internal and external communications, and monitor organizational performance against strategic goals. It stands as an important framework that can be applied to measure the performance of a CoE. It encompasses diverse perspectives, including financial, customer, internal processes, and learning and growth dimensions. Although a CoE might not directly generate revenue, you can measure its cost-effectiveness or how it helps other parts of the organization save costs or improve revenue. For a CoE, "customers" can include internal stakeholders or business units that depend on the expertise and outputs of the CoE. Metrics can include satisfaction rates, the usability of the solutions provided, and the degree of integration of CoE initiatives. Internal processes can focus on the efficiency and effectiveness of the CoE's processes, such as the speed of developing new solutions, the quality of knowledge created, and the success rate of implemented projects, and finally, the learning and growth perspective is crucial for a CoE as it revolves around knowledge and expertise. Metrics can include employee skill advancements, innovation rates, and improvements in tools and methodologies [5].

For customized CoEs that reflect industry-specific subtleties, such as environmental sustainability and regulatory compliance, the Balanced Scorecard enables CoEs to align their KPIs with the overarching strategic goals of the organization. Leveraging this framework enables CoEs to gain a holistic understanding of their impact across multiple dimensions, facilitating strategic alignment and informed decision-making.

ROI analysis is another useful tool for quantifying the financial returns generated by CoE initiatives. Through ROI analysis as described in the previous section, CoEs can measure the tangible benefits derived from their activities, be it cost savings, or efficiency improvements against the corresponding investment costs. This financial metric provides a clear and tangible measure of CoE effectiveness, enabling organizations to gauge the return on their investment and make informed resource allocation decisions.

Benchmarking represents a valuable practice for evaluating CoE performance in relation to industry benchmarks or competitors. Benchmarking offers invaluable insights into performance gaps, industry best practices, and opportunities for improvement, comparing their performance against industry benchmarks to identify areas of strength and areas needing improvement, therefore guiding targeted interventions and strategic initiatives to enhance overall effectiveness.

Establishing clear KPIs

Defining appropriate KPIs is one of the foundational steps in measuring CoE success. In the energy sector, where CoEs are instrumental in driving innovation, sustainability, and efficiency, KPIs must be well-defined and tailored to align with specific objectives. Here is a more in-depth look at some potential KPIs to measure CoE performance.

- Innovation adoption rates

    For CoEs primarily focused on fostering innovation, measuring innovation adoption rates is crucial. This KPI assesses how quickly new technologies, processes, or practices developed by the CoE are integrated into the organization's operations. High adoption rates indicate that the CoE is successful in influencing the organization's approach to innovation.

- Efficiency improvements

    CoEs dedicated to enhancing operational efficiency should track metrics related to cost reduction and process optimization. Specific KPIs could include the percentage reduction in operational costs, cycle time improvements, or resource utilization efficiency. These KPIs directly reflect the CoE's impact on optimizing the energy organization's processes.

- Sustainability targets achieved

    In an era of heightened environmental consciousness, CoEs focusing on sustainability initiatives should measure their success by tracking progress toward specific sustainability goals. KPIs might include reductions in carbon emissions, increases in the share of renewable energy sources, or improvements in resource conservation. Achieving or surpassing these targets demonstrates the CoE's commitment to sustainability.

- Customer satisfaction scores

   For CoEs influencing customer-facing aspects of the business, measuring customer satisfaction is essential. This KPI captures the impact of CoE-driven initiatives on customer experiences. Higher customer satisfaction scores indicate that CoE efforts are positively influencing how customers perceive and interact with the energy company.

- Cost savings and ROI

   CoEs with a cost optimization focus should calculate and demonstrate cost savings realized through their initiatives. This KPI involves comparing the cost reductions achieved against the investments made in CoE activities. Positive ROI is a clear indicator of the CoE's financial impact and value.

CoE challenges and pitfalls

A CoE is a powerful vehicle for driving innovation, fostering collaboration, and achieving organizational excellence. However, despite its immense potential CoEs are not immune to challenges or pitfalls that can impede their effectiveness and hinder their success. The following section explores some common pitfalls faced by CoEs and discuss strategies for overcoming these challenges to ensure their continued relevance and impact.

Strategic misalignment and lack of direction

A fundamental challenge for CoEs is the absence of a coherent strategy and its misalignment with the organizational vision. This misalignment can lead to difficulties in delivering value and fostering innovation, as CoEs might not fully grasp or contribute toward the organization's goals and objectives. It is crucial for CoEs to integrate their initiatives with the organization's strategic priorities, ensuring that their efforts contribute to significant and meaningful achievements.

Ambiguity in leadership and governance

The lack of a defined leadership and governance structure within CoEs often leads to confusion and inefficiency. Ambiguity in leadership roles between the CoE and other organizational units can cause coordination issues and hinder the execution of initiatives. Visible leadership with buy in from all stakeholders is important to ensure the organization gets behind the CoE and supports it from its early establishment.

Knowledge silos and limited sharing

A tendency toward insularity within CoEs can severely restrict the flow of knowledge and expertise across the organization. When knowledge remains siloed, it stifles innovation and collaboration preventing the organization from accessing a wide range of insights and capabilities. CoEs must foster a culture of openness and knowledge sharing, encouraging cross-functional interactions and collaboration to overcome these barriers.

Inadequate performance measurement

The absence of a comprehensive framework for assessing performance can leave CoEs unable to prove their worth and impact. Developing clear metrics and measurement tools as presented in the previous sections is vitally important to evaluate the success of CoE initiatives and to pinpoint areas needing improvement. Through regular assessments, CoEs can monitor their progress, align their outcomes with set benchmarks, and make informed decisions to enhance their operations.

Narrow focus and lack of transformational impact

Focusing solely on minor improvements can trap CoEs in a cycle of incremental change, preventing them from pursuing more significant, transformative initiatives. While continual enhancements are key, CoEs should also aim for strategic, high-impact projects that can lead to substantial organizational shifts. A broad operational scope enables CoEs to stay flexible and responsive to changing business needs, ensuring the organization remains competitive.

Organizational resistance to change

Internal resistance within the organization can impede the adoption of new practices and innovations introduced by CoEs. This resistance may arise from a variety of sources, including entrenched organizational cultures, fear of the unknown, or concerns over job security. To overcome this, CoEs must employ effective communication, engage stakeholders at all levels, and apply comprehensive change management practices to mitigate resistance and foster a culture of adaptability.

Constraints on resources

Resource limitations, whether in terms of funding, talent, or technology, can significantly reduce the effectiveness of CoEs. Insufficient resources may prevent CoEs from fully implementing their initiatives, stifling innovation and organizational growth. It is imperative for organizations to commit to supporting their CoEs, providing them with the necessary resources to make a meaningful impact.

Lack of comprehensive stakeholder engagement

Insufficient engagement with key stakeholders, including senior leaders, frontline staff, and external partners, can undermine the CoEs' efforts. Engaging stakeholders is essential for securing support, encouraging collaborative efforts, and ensuring the successful execution of CoE initiatives. CoEs should actively involve stakeholders in all stages, from strategy formulation to implementation and evaluation, to ensure alignment and foster a sense of shared purpose.

Addressing these common pitfalls requires proactive measures, including clear communication, effective leadership, robust performance evaluation mechanisms, and a commitment to continuous improvement. Overcoming these challenges and put CoEs on the right path to add real value and maximize their impact on organizations.

How to setup a CoE

Establishing a CoE is a strategic initiative that necessitates a detailed and adaptable framework to guide its development and operation, which is key for aligning the CoE's activities with the overarching goals of the organization, particularly in an industry as dynamic and critical as energy.

Setting up a CoE requires a structured approach that aligns with the organization's strategic goals and industry challenges. Each step of the following proposed framework is designed to establish a foundation, provide direction, and ensure the CoE's initiatives contribute to the organization's overarching objectives. The following framework provides a comprehensive approach to setting up a CoE, tailored to meet the distinctive challenges and opportunities in the energy sector.

1. Define a CoE vision

    The vision of a CoE serve as its guiding star, clearly articulating the purpose, ambitions, objectives, and expected outcomes of the initiative.

This step is fundamental for setting the direction and scope of the CoE, ensuring that its efforts are aligned with the broader objectives of the organization. A well-defined mission and vision statement help in communicating the CoE's goals to all stakeholders, fostering a shared understanding and commitment to the CoE's strategic role within the organization.

The first step involves creating clear vision for the CoE that encapsulates its purpose, aspirations, and the specific objectives and outcomes it aims to achieve. This should resonate with the broader objectives of the energy company, especially in areas, such as promoting renewable energy sources, enhancing efficiency, and integrating innovative technologies. The CoE's vision should align with the energy company's strategic commitments be it innovation, environmental sustainability, operational excellence, etc. This alignment ensures that the CoE's efforts contribute directly to the company's long-term success and sustainability initiatives.

2. Governance structure

Establishing a governance structure is fundamental to a CoE's success, as it defines how decisions are made, responsibilities are allocated, and performance is monitored. This step involves setting up a structure that ensures accountability, transparency, and effective leadership across the CoE's operations. A clear governance structure supports the CoE in achieving its objectives by facilitating efficient decision-making, enhancing organizational alignment, and ensuring that the CoE's activities are closely monitored and aligned with the company's strategic priorities.

A well-defined governance structure is critical for the CoE's success, providing the necessary oversight, accountability, and direction. This structure should delineate the roles and responsibilities of all involved parties, from executive sponsors to operational teams, ensuring clear lines of authority and decision-making processes.

The governance structure should also implement mechanisms for regular and transparent reporting on the CoE's activities, achievements, and challenges. This ensures that all stakeholders are informed and can contribute to the CoE's continuous improvement.

3. Core functions

The core functions of a CoE define its primary roles and responsibilities within the organization, including innovation management, knowledge sharing, project support, and performance measurement. This step is about identifying and prioritizing the activities that will deliver the most value to the organization, ensuring that the CoE effectively drives innovation, fosters a culture of continuous learning, and supports the successful execution of strategic initiatives. It is important to clearly define the CoE's core functions, so that the CoE can focus its resources and efforts on areas where it can have the greatest impact.

Innovation management is a core function of the CoE, aimed at identifying and leveraging emerging technologies and business models to propel the energy company forward. This involves fostering an environment where creativity and experimentation are encouraged and valued. The CoE should also serve as a hub for knowledge sharing, capturing valuable insights, and disseminating them throughout the organization. Additionally, it should offer project management expertise to support and guide strategic initiatives, ensuring they are executed efficiently and effectively.

4. Resource allocation

Resource allocation is a critical step in the framework, ensuring that the CoE has the necessary funding, personnel, and technology to achieve its vision and objectives. This step involves strategically distributing resources to support the CoE's operations and initiatives, prioritizing investments that align with the CoE's mission and the organization's strategic goals. Effective resource allocation is key to enabling the CoE to execute its functions efficiently and drive meaningful innovation and improvement within the company.

The allocation of resources, including budget, personnel, and technology, is critical for enabling the CoE to fulfill its mission. Resources should be allocated strategically, prioritizing initiatives that align with the CoE's and organization's strategic objectives.

5. Stakeholder engagement

Stakeholder engagement is essential for the success of a CoE, as it ensures that leadership are bought in to the CoE, the CoE initiatives are well supported, aligned with business needs, and effectively implemented across the organization. This step focuses on identifying, communicating with, and involving key stakeholders from within and outside the organization. Fostering strong relationships with stakeholders enables the CoE to secure the necessary buy-in, collaboration, and support to successfully carry out its initiatives and drive organizational change.

Engagement with a wide range of stakeholders, both internal and external to the organization, is important for the CoE's success. This engagement fosters collaboration and ensures that CoE is supported and aligned with the needs and goals of the broader organization.

6. Fostering knowledge sharing and collaboration

Promoting knowledge sharing and collaboration within the CoE and across the organization is vital for leveraging collective expertise and driving innovation. This step is about creating an environment and platforms that encourage the exchange of ideas, best practices, and lessons learned. By facilitating collaboration and knowledge sharing, the CoE can enhance organizational learning, improve efficiency, and foster a culture that supports continuous improvement and innovation.

The CoE may establish forums and platforms to facilitate knowledge exchange and collaboration, breaking down silos and promoting cross-functional teamwork. This includes supporting communities of practice and organizing learning opportunities to enhance the skills and knowledge of CoE members and other employees.

7. Performance evaluation and reporting

Evaluating performance and reporting on the CoE's achievements and challenges are crucial for demonstrating value and guiding future initiatives. This step focuses on developing a comprehensive framework for assessing the CoE's impact, utilizing both qualitative and quantitative measures. Regular performance evaluation and transparent reporting help in maintaining accountability, informing stakeholders about the CoE's progress, and identifying areas for further development and improvement.

Developing and implementing KPIs and metrics to evaluate the CoE's initiatives is essential for assessing their impact and guiding decision-making. This also facilitates a culture of data-driven improvement and accountability. A well-defined and clear purpose serves as the foundation for a successful CoE, providing transparency, structure, and direction for its operations.

8. Emphasizing continuous improvement

Continuous improvement is a key principle for CoEs, enabling them to adapt to changing circumstances, optimize operations, and enhance their impact over time. This step involves establishing processes for regular review, feedback, and refinement of the CoE's activities and strategies. Embedding continuous improvement into the CoE's culture permits the organization to remain agile, responsive to new challenges and opportunities, and committed to achieving excellence in its operations.

Establishing mechanisms for ongoing improvement, leveraging feedback and evaluations to refine and enhance CoE operations is an important part of the implementation of a CoE. This approach promotes a culture of learning and adaptation, encouraging team members to innovate and embrace new challenges.

Each of these steps contributes to a robust and effective CoE, ensuring that it is well-positioned to support the organization's strategic objectives, drive innovation, and enhance performance.

Implementation

The implementation of a CoE is a strategic endeavor that requires a structured approach to ensure its success with buy in at all levels in the organization. The implementation process typically involves several stages, each of which plays a crucial role in laying the foundation, establishing operations, and driving value. Below is an overview of the key stages involved in implementing a CoE (Fig. 6.1):

1. Initial planning and assessment

   To initiate the establishment of a CoE, an organization should begin by conducting a comprehensive assessment of its current state, identifying strengths, weaknesses, opportunities, and threats that are relevant to the new initiative. This evaluation will provide a clear foundation for defining the scope, objectives, and expected outcomes of the CoE, ensuring that they align with the strategic priorities and goals of the energy company. Following this groundwork, it is important to establish a project team composed of key stakeholders, subject matter experts, and project managers. This team will oversee the implementation process and guide decision-making, playing a pivotal role in the successful realization of the CoE.

2. Strategy development

   In the strategy development phase for the CoE, it is important to formulate a clear and coherent strategy that encompasses the mission, vision, goals, and objectives of the CoE, aligning them with the organization's overarching strategy. This includes defining the target areas of focus for the CoE, such as innovation, operational excellence, sustainability, or digital transformation, which should be selected based on the identified needs and strategic priorities of the organization.

   Additionally, determining the governance structure, resource requirements, and performance metrics is essential to support the successful operation and accountability of the CoE. These elements collectively ensure that the CoE is strategically positioned to drive meaningful change and add value to the organization.

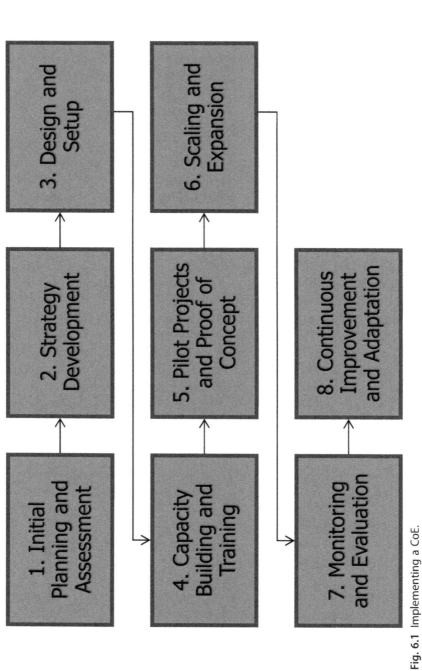

Fig. 6.1 Implementing a CoE.

3. Design and setup

During the design and setup phase of the CoE, a keen attention to detail in the design of the organizational structure clearly specifying roles, responsibilities, reporting lines, and decision-making processes is key to ensure effective governance and smooth operations.

Additionally, establishing both the physical and virtual infrastructure is fundamental to support the activities of the CoE. This includes setting up appropriate office spaces, technology platforms, communication channels, and collaboration tools that facilitate efficient workflow and interaction among team members.

The process also involves recruiting and onboarding the core team members of the CoE, including leadership, subject matter experts, project managers, and support staff. It is important to ensure a diverse mix of skills and expertise within the team to foster innovative solutions and robust project execution.

4. Capacity building and training

In the capacity building and training phase of the CoE, it is important to provide ample training and development opportunities for CoE members to enhance their knowledge, skills, and capabilities in areas, such as technology, project management, innovation, and change management. This investment in human capital not only boosts individual competencies but also contributes to the overall effectiveness of the CoE.

Simultaneously, fostering a culture of continuous learning, collaboration, and knowledge sharing within the CoE is vital. Encouraging team members to stay informed about industry trends, best practices, and emerging technologies promotes a dynamic and innovative environment where collaborative efforts and knowledge exchange are commonplace, thereby driving the CoE toward achieving its strategic goals.

5. Pilot projects and proof of concept

Launching pilot projects or proof-of-concept initiatives is a key step in the establishment of a CoE to test its effectiveness for the organization. These initial projects serve as practical experiments that allow the CoE to demonstrate the viability of its ideas and strategies. Implementing these controlled, smaller-scale projects, allows the CoE to observe how its theories work in practice and how it is adding value to the organization, allowing for a better understanding of the potential impacts and effectiveness of its initiatives.

Following the initial roll-out of pilot projects, gathering feedback to evaluate the CoEs performance and iterate on these projects is important to fine tune the CoE. This process helps to refine approaches, address any challenges encountered, and optimize outcomes. The insights gained from these evaluations are invaluable for making necessary adjustments before considering a broader implementation of CoE activities. This iterative process ensures that the strategies and solutions developed are not only effective and efficient but also aligned with the overall goals and capabilities of the organization, setting the stage for successful scaling up of the CoE's initiatives.

6. Scaling and expansion

As the CoE demonstrates success and maturity in its initial projects, the next strategic step involves gradually scaling up the scope and impact of its activities. This expansion can involve extending CoE's influence into additional areas of the organization or embarking on larger and more ambitious projects and initiatives. The goal of scaling is to maximize the reach and effectiveness of the CoE, ensuring that its benefits are disseminated widely throughout the organization and that its solutions are applied to a broader spectrum of challenges and opportunities.

To support this growth, it is important to foster partnerships and collaboration with other departments, business units, and external stakeholders. Leveraging these relationships can provide the CoE with access to additional resources, specialized expertise, and essential support, all of which are vital for the successful execution and enhancement of CoE initiatives. These collaborative efforts can enhance the capabilities of the CoE and help embed its practices and principles across the organization, driving a more integrated and comprehensive approach to innovation and excellence.

7. Monitoring and evaluation

To ensure the success and relevance of the CoE, it is advisable to establish robust mechanisms for monitoring and evaluating its performance, impact, and value. These mechanisms should compare the outcomes of the CoE's activities against predefined objectives and KPIs. This systematic evaluation helps in assessing whether the CoE is meeting its set goals and contributing effectively to the organization's broader objectives. Implementing these monitoring processes can help the organization to maintain a clear view of the CoE's progress and its contributions to strategic targets.

Furthermore, conducting regular reviews of the CoE's progress by analyzing data, and soliciting feedback from stakeholders is a crucial part of the evaluation step in setting up a CoE. This ongoing evaluation process allows for the identification of areas that require improvement, facilitates timely course corrections, and ensures that the CoE's activities remain aligned with the evolving goals of the organization. Gathering insights from a broad range of stakeholders not only enriches the feedback but also supports the CoE in making informed decisions that enhance its effectiveness and sustainability. This iterative review and refinement process fosters a culture of continuous improvement within the CoE, ensuring its long-term success and relevance.

8. Continuous improvement and adaptation

Fostering a culture of continuous improvement and adaptation within the CoE is vital for maintaining its relevance and efficacy in the dynamic business environment. Encouraging innovation, experimentation, and agility within the CoE helps it respond proactively to changing business needs and market dynamics. This culture empowers team members to think creatively, challenge the status quo, and explore new ways to enhance organizational processes and outcomes. Nurturing an environment that welcomes change and quick adaptation enables the CoE to remain flexible and effective, and ready to tackle new challenges as they arise.

Embracing feedback, lessons learned, and best practices is essential for the ongoing refinement and enhancement of the CoE's operations. Regularly integrating new insights allows the CoE to continuously improve its processes and capabilities, thereby driving sustainable results over time. This approach ensures that the CoE adapts to current conditions and anticipates future changes, thereby maintaining its strategic edge. Consistently applying these learned insights and best practices helps the CoE to enhance its operational efficiency and effectiveness, contributing significantly to the organization's long-term success.

The implementation of a CoE in the organization is a multifaceted process that requires strategic planning, meticulous execution, and ongoing refinement. It is important to follow a structured approach and leverage the collective expertise and commitment of key stakeholders to establish CoEs that can drive innovation, foster collaboration, and deliver tangible value to the organization and its stakeholders.

References

[1] C.S. Pemberton, What Makes a Marketing Center of Excellence? Harvard Business Review, 2016. https://hbr.org.
[2] GCC.SG.Org, GCC Master Series, 2020.

[3] CEO Innovation Centre, 2022. https://www.aramco.com/.

[4] W.E. Deming, Quality, Productivity, and Competitive Position, Massachusetts Institute of Technology, Center for Advanced Engineering Study, 1982.

[5] R.S. Kaplan, D.P. Norton, Using the Balanced Scorecard as a Strategic Management System, Harvard Business Review, 2007.

CHAPTER SEVEN

Delivery Assurance

The energy sector stands at the cusp of a monumental shift, propelled by the global imperative for a sustainable and secure energy future. This transition, marked by the integration of renewable energy sources, advancements in technology, and evolving regulatory landscapes, presents a realm of opportunities and challenges. Central to navigating this transition successfully is the sector's ability to deliver projects that are not only innovative but also resilient to the uncertainties inherent in this evolution. The dynamic nature of the energy transition with many project scopes continuously evolving and technologies still in development accentuates the need for a robust, structured, yet agile approach to project delivery.

This agility is crucial as it allows energy companies to adapt to changing circumstances, incorporate new technologies, and respond to unforeseen challenges without compromising on project outcomes. The complexity and scale of projects in the energy sector further drive the need for delivery certainty. Traditional project management methodologies, while still relevant, must evolve to incorporate more collaborative contracting approaches, emphasizing partnership and shared objectives over transactional relationships. Additionally, leveraging digital and data-driven approaches has become indispensable, offering unprecedented insights into project performance, risk management, and decision-making.

Given these reflections, "Delivery Assurance" emerges as a pivotal process within project management frameworks in the energy sector. It encompasses a set of practices and principles designed to ensure projects are delivered on time, within budget, and in alignment with the strategic goals of the organization and the broader imperatives of the energy transition. This chapter delves into the importance of a structured, yet agile approach to project delivery in the energy sector, exploring how it underpins the success of the energy transition.

In this chapter, the principles presented in Chapter 3, the role of collaborative contracting, the impact of digitalization from Chapter 4, and deployment of a data-driven principles from Chapter 5, will be drawn upon and come together to address the critical nature of Project Delivery Assurance in the energy transition.

Powering Through the Transition
https://doi.org/10.1016/B978-0-323-91754-4.00006-6

Challenges and opportunities in project delivery

The energy sector is undergoing an unprecedented transformation, driven by the burning need for a more sustainable, reliable, and clean energy supply. This energy transition is characterized by the shift from traditional fossil fuels to renewable energy sources, the integration of innovative technologies, and the adaptation to changing regulatory environments and market dynamics. As the sector evolves, the complexity and scope of energy projects increase, as emphasized by the uncertain and dynamic nature of the transition itself. In this context, the ability to deliver projects effectively, balancing innovation with reliability and sustainability becomes a critical capability for energy companies.

The inherent uncertainties of the energy transition demand a project delivery approach that is both structured and agile. Traditional project management methodologies, while providing a solid foundation, often lack the flexibility to adapt to rapid changes or to incorporate new insights and technologies as projects evolve. This is particularly relevant as many projects in the energy sector are pioneering in nature, involving technologies or processes that are still in development or being scaled up for the first time.

Agile project delivery, in this context, does not merely refer to the use of agile methodologies often associated with software development but to a broader philosophy of flexibility, continuous improvement, and adaptability. It involves iterative planning and development, where project scopes are allowed to evolve based on learning and discoveries made during the project lifecycle. This approach enables energy companies to respond to new opportunities and challenges more effectively, ensuring that projects remain aligned with the ultimate goals of sustainability and efficiency.

Furthermore, the complexity and scale of energy transition projects necessitate a focus on "Delivery Assurance," a comprehensive process designed to ensure that projects are not only delivered within time and budget constraints but also achieve the intended outcomes and value. Delivery Assurance encompasses a wide range of practices, from rigorous risk management and Quality Assurance (QA) to stakeholder engagement and performance monitoring. It acts as a safeguard, ensuring that despite the agile and iterative approach to project delivery, the end goals of the project are never compromised.

The adoption of more collaborative contracting approaches marks another shift in the project delivery landscape. Moving away from traditional, transactional contract models, energy companies are now embracing partnerships and alliances that foster mutual trust, share risks and rewards, and align objectives among all project stakeholders. This collaborative mindset is crucial for tackling the complexities and uncertainties of energy transition projects, where the integration of diverse technologies, disciplines, and expertise is essential for success.

Moreover, digital and data-driven approaches are increasingly becoming integral to project delivery in the energy sector. The utilization of advanced analytics, digital twins, and project management software offers unprecedented visibility into project performance, risks, and opportunities. These tools not only enhance decision-making and efficiency but also play a vital role in supporting the agile and adaptive project management strategies required for the energy transition.

As the energy sector continues to navigate through this period of change, adopting the principles of agile project delivery, underpinned by a strong focus on Delivery Assurance, collaborative contracting, and digitalization, is an imperative for success. These approaches offer a pathway to managing the complexities and uncertainties of the energy transition, ensuring that projects can be delivered with certainty and contribute effectively to the global shift toward a more sustainable and resilient energy future.

One of the fundamental prerequisites of an effective Delivery Assurance Program is the concept of *Project Criticality*, which in basic terms enables prioritization of "critical" projects to more efficient and effective in the deployment of the organizations resources.

Criticality

The concept of criticality is a fundamental principle that has been around since the early 20th century and permeates various fields and disciplines. In basic terms, it highlights the importance or urgency of specific elements or tasks within a system or process. It is a common concept, yet it carries immense significance as it often follows Pareto's principle, also known as the 80/20 rule. This principle suggests that approximately 80% of effects come from 20% of the causes. In the context of project management and delivery, criticality helps identify the 20% of projects that are likely to have the most substantial impact on an organization's performance and success. The concept of criticality in project management serves as a pivotal mechanism for assessing and prioritizing the potential impact of risks across various projects, particularly within high-stakes industries like energy.

It involves systematically evaluating projects to identify those that carry the greatest threat of negative outcomes should risks materialize, such as significant cost overruns, schedule delays, or failure to meet regulatory and safety standards. This evaluation is grounded in a thorough analysis of risk probability and impact, enabling organizations to proactively allocate resources, focus management efforts, and apply tailored risk mitigation strategies. As such, criticality acts as an essential filter in project delivery, ensuring that the most significant projects, those upon which organizational success heavily relies are managed with the diligence and oversight necessary for effective and efficient completion.

Problems can frequently occur on projects, and if left unaddressed can snowball into costly failures. Recognizing the signs of trouble as soon as possible is crucial for timely intervention and mitigation. The project criticality approach acts as an early warning system, flagging projects that may be more susceptible to challenges and providing additional assurance measures. Costly project failures are not only financially burdensome but can also erode an organization's reputation and stakeholder trust. The criticality assessment process acts as a shield against such failures.

Project criticality is a systematic process designed to gauge the relative importance of each project in an organization's portfolio. Implementing a project criticality approach provides a structured early warning system. It flags projects that are more susceptible to challenges due to their complexity, scope, importance, or external factors. Such projects receive additional assurance measures, which may include increased monitoring, more frequent reviews, specialized risk management strategies, and the application of best practices in project management. This vigilance is designed to reduce the likelihood of setbacks and ensure that projects progress as planned, or when deviations occur, they are managed with minimal disruption.

Project criticality approaches

For companies that handle a large number of projects simultaneously, the project criticality process becomes a key component of a wider Delivery Assurance strategy. It aids in prioritizing high-risk projects, thereby guiding the allocation of limited overhead resources in the most efficient way possible. Early identification of problems within these high-priority projects allows for proactive measures, significantly reducing the likelihood of project failures or setbacks.

It is important to be clear about your categorization of project risks, since the term risk management is relevant to its context, e.g., project complexity, commercial model, and size of project. In the energy sector, project services companies can adopt a criticality methodology that combines industry-specific considerations with proven project management practices. This approach would allow these companies to prioritize projects based on risk, complexity, strategic importance, and potential impact. Here are some approaches that can be integrated to develop a criticality methodology.

1. Risk-based prioritization

    Projects could be evaluated based on the level of risk they pose to the company, including technical, financial, environmental, and compliance risks among many others. A risk matrix can be drawn up and utilized, scoring projects based on the likelihood of risk occurrence and the severity of its impact. This approach helps in identifying high-risk projects that need closer attention and resources.

2. Strategic alignment

    Projects could be assessed for their alignment with the company's strategic objectives, such as expanding into new markets, improving sustainability, or adopting new technologies. Projects that are closely aligned with these strategic goals might be deemed more critical.

3. Financial impact

    Projects could be analyzed for their financial impact, considering the potential revenue, margin, and the consequences of budget overruns or delays. Projects with a higher financial stake or those critical for the financial health of the company should be prioritized.

4. Stakeholder impact assessment

    Identifying and evaluating the influence of stakeholders on the project and vice versa can determine the project's criticality. Projects with high stakeholder interest or those that significantly impact key stakeholders, including the public or regulatory agencies, often warrant higher criticality.

5. Regulatory and compliance relevance

    Given the regulatory intensity in the energy sector, projects could be evaluated on their regulatory and compliance demands. Projects with significant regulatory hurdles or those contributing to regulatory compliance could rank higher in criticality.

6. Complexity and technical difficulty

    Evaluate projects based on the complexity of the work involved and the technical challenges they present. Projects with higher levels of complexity and technical difficulty, which also carry a higher risk of failure, should be prioritized for more rigorous management and oversight.

7. Schedule sensitivity analysis

Determine the criticality of projects based on their schedule sensitivity. Projects on the critical path that could impact the delivery of other projects or those with fixed statutory or contractual deadlines should be considered more critical.

8. Historical data and lessons learned

Leverage historical project data and lessons learned to inform the criticality assessment. Projects similar to past ones with issues should be flagged for additional scrutiny.

9. Environmental and social governance (ESG) factors

Projects should be evaluated for their impact on ESG factors, which are increasingly important in the energy sector. Projects that contribute to the company's ESG goals may be considered more critical in terms of reputation and long-term sustainability.

As you can see, there are numerous options that could be explored to determine project criticality and the list above is not exhaustive. It depends on the specifics of the organization in question as to which one or a combination fits best. Managing these criticality assessments would require a robust project management information system that allows for the tracking of these various factors and provides a platform for scoring and ranking projects. This system should facilitate easy updating of scores as projects progress or as conditions change, ensuring that the criticality assessment remains a living part of the project management process. This is where adopting a data-driven approach comes into play as outlined in Chapter 5.

It is important to note that these approaches should be tailored to the specific needs and circumstances of the company and the projects it undertakes. Additionally, industry standards and methodologies, such as the Project Management Institute's (PMI's) Project Management Body of Knowledge, should be considered when developing these criticality assessment methods to ensure that they are comprehensive and adhere to best practices.

Project delivery assurance

Delivery Assurance is a sophisticated and fundamental concept that ensures projects are not just executed but are executed in a manner that shifts away from mere project completion to a holistic achievement that encompasses timely delivery, budget adherence, quality standards, and strategic alignment with the organization's goals and the broader vision of the energy transition. This shift is crucial in an industry characterized by rapid technological advancements, regulatory shifts, and an urgent need for sustainability. Delivery Assurance demands an in-depth understanding of these elements, integrating a wide array of practices, including agile project management, stakeholder engagement, data driven, and continuous innovation. The goal is to navigate the project through the complexities of the modern energy landscape, ensuring it leverages the latest technologies, complies with current regulations, and meets the evolving demands of stakeholders. By doing so, Delivery Assurance aims to achieve project objectives and drive the organization forward in the competitive and ever-changing energy market.

Achieving Delivery Assurance, however, presents its own set of challenges, requiring a balance between flexibility and control. The dynamic nature of the energy sector, with its unpredictable technological and regulatory changes, calls for a project management approach that is both adaptive and robust. Effective Delivery Assurance frameworks thus incorporate elements like real-time risk management, iterative planning (agile project delivery), and stakeholder engagement to ensure projects can adapt to changes without losing sight of their strategic goals. These frameworks also emphasize the importance of aligning projects with global sustainability targets, making Delivery Assurance not just a method for project success but a contribution to the larger goal of energy transition. While challenging, the successful implementation of Delivery Assurance is achievable through strong leadership, a culture of continuous improvement, and the use of technology to enhance project visibility and decision-making. In doing so, organizations can ensure that their projects are not only completed successfully but also contribute meaningfully to the sustainability and resilience of the energy sector.

Project Delivery Assurance incorporates several key elements designed to ensure that projects meet their objectives despite the inherent challenges. These elements are crucial for navigating the intricacies and uncertainties of complex projects coupled with the challenging demands of the energy transition, ensuring alignment with strategic goals, and achieving successful outcomes.

Here are some of the key elements of Project Delivery Assurance
1. Scope definition and comprehensive planning
Detailed planning is fundamental to the success of complex projects. This involves defining the project scope clearly, establishing realistic timelines, and allocating resources efficiently. Comprehensive planning also includes contingency planning to address potential risks and challenges that may arise during project execution.

Scope definition and comprehensive planning serve as the cornerstone of Project Delivery Assurance for complex projects. This initial phase lays the groundwork for project success by meticulously detailing what the project aims to achieve and outlining the path to get there. It involves several critical steps, each contributing to a thorough understanding and preparation for the project's demands, opportunities, and constraints.

Firstly, clearly defining the project scope is paramount. This step involves identifying the specific goals, deliverables, and functions of the project, as well as understanding the needs and expectations of stakeholders. A well-defined scope sets clear boundaries for what the project will and will not include, helping to prevent scope creep and ensuring that the project team remains focused on the agreed objectives.

Effective resource allocation is essential in comprehensive planning. This involves assigning the right mix of personnel, budget, and technology resources to various project tasks. By carefully planning how resources are allocated, project managers can ensure that the project team has everything they need to succeed, from skilled personnel to necessary equipment and funding.

Establishing realistic timelines is another critical aspect of comprehensive planning. This includes setting milestones and deadlines for key project phases and deliverables. Developing a timeline helps in creating a roadmap for the project, enabling the team to track progress and make timely adjustments as needed. It also aids in setting expectations with stakeholders regarding project duration and key delivery points.

Given the uncertainties inherent in complex projects, contingency planning is a crucial part of the planning process. This involves anticipating potential risks and challenges that could impact the project and developing strategies to mitigate these risks. Contingency plans ensure that the project team is prepared to address problems without significant delays or cost overruns.

2. Stakeholder engagement and communication

Effective stakeholder engagement ensures that all parties involved are aligned with the project's objectives and expectations. Regular, transparent communication keeps stakeholders informed about project progress, changes, and challenges, facilitating timely decision-making and support.

Effective stakeholder engagement and communication are pivotal in the successful delivery of complex projects, serving as a vital component of Project Delivery Assurance. This involves a strategic approach to identifying, understanding, and managing the relationships with all parties impacted by the project or who can impact the project. By focusing on this element, project managers can build a collaborative environment that fosters support, mitigates conflicts, and aligns expectations, thereby enhancing the project's chances for success.

The first step in stakeholder engagement is to identify who the stakeholders are, ranging from internal teams and management to external clients, suppliers, regulators, and the community. Understanding their interests, expectations, influence, and potential impact on the project is crucial. This analysis helps in prioritizing stakeholders and tailoring engagement strategies to address their specific needs and concerns.

A comprehensive communication plan is essential for effective stakeholder engagement. This plan should outline the frequency, methods, and formats of communication tailored to the needs of different stakeholder groups. Whether it is regular status updates, steering committee meetings, or ad hoc communications to address specific issues, the goal is to ensure that information flows effectively in both directions, providing stakeholders with the information they need while also gathering their feedback and input.

Successful stakeholder engagement is rooted in building strong relationships and trust. This involves being transparent about project goals, progress, challenges, and acknowledging stakeholder concerns. Demonstrating a commitment to addressing these concerns and incorporating stakeholder feedback into the project where appropriate allows project managers to instil a sense of ownership and support among stakeholders.

One of the key challenges in stakeholder engagement is managing expectations. This includes setting realistic expectations about project outcomes, timelines, and potential risks from the outset. Regular updates and honest communication about project status and any changes are vital to keeping stakeholders informed and managing their expectations throughout the project lifecycle.

Effective stakeholder engagement is an iterative process that benefits from establishing feedback loops. Encouraging and acting on feedback from stakeholders helps not only in making informed decisions but also in adapting project strategies as necessary. This adaptability can lead to improvements in project processes, outcomes, and stakeholder satisfaction.

Focusing on stakeholder engagement and communication ensures that complex projects are conducted in an environment where transparency, collaboration, and mutual respect prevail. This approach helps in navigating the challenges inherent in complex projects and also ensures that the project remains aligned with the needs and expectations of those it impacts, contributing significantly to its successful delivery.

3. Risk management

Identifying, assessing, and mitigating risks are crucial in complex projects. A proactive approach to risk management includes continuous monitoring of potential threats to the project's success and implementing strategies to minimize their impact. This also involves quality control measures to ensure that the project's deliverables meet the required standards.

Risk management and quality control are central to the concept of Project Delivery Assurance, particularly in the context of complex projects. This dual focus ensures that potential obstacles to project success are identified early and managed proactively, while the quality of deliverables meets or exceeds the project's requirements and stakeholders' expectations.

At the heart of effective risk management is the identification, assessment, and prioritization of potential risks that could negatively impact the project. This process begins with a comprehensive risk identification exercise, utilizing tools, and techniques, such as brainstorming sessions, SWOT analysis (strengths, weaknesses, opportunities, threats), and expert reviews to compile a broad list of potential risks. Following identification, each risk is assessed for its likelihood of occurrence and potential impact on the project. This assessment allows project managers to prioritize risks, focusing attention and resources on managing those that pose the greatest threat to project success.

Mitigation strategies are then developed for significant risks, tailored to either reduce the likelihood of the risk occurring or to minimize its impact should it materialize. This proactive approach to risk management is not a one-time activity but a continuous process that requires regular monitoring and reassessment of risks throughout the project lifecycle. By staying vigilant and responsive, project teams can adapt their strategies as project conditions change, ensuring resilience and flexibility in the face of uncertainty.

4. Agile and adaptive methodologies

Implementing agile and adaptive methodologies allows for flexibility in project execution. This approach accommodates changes in project scope, requirements, and external factors, enabling the project team to respond effectively to unforeseen challenges without significantly derailing the project.

Agile project management is a central subset of the broader concept of Project Delivery Assurance, embodying a flexible, iterative approach that is especially suited to managing the complexities and rapid changes often encountered in project environments and especially suited to complex energy transition projects. At its core, agile focuses on delivering value through collaborative efforts, adaptive planning, and continuous improvement, principles that align closely with the goals of ensuring project success and alignment with strategic objectives. This methodology integrates with the overarching framework of Delivery Assurance emphasizing responsiveness to change, stakeholder engagement, and the delivery of high-quality outcomes in a timely manner.

Agile project delivery, particularly within the evolving landscape of the energy sector, represents a strategic pivot from traditional project management methodologies to a more flexible, responsive, and iterative approach. This transition is driven by the sector's inherent uncertainties, including evolving regulatory policies, technological advancements, and shifting market demands. The key pillars of agile project delivery include flexibility, adaptability, and continuous improvement serve as foundational elements that enable energy companies to navigate these uncertainties successfully [1].

Flexibility and adaptability are crucial for managing the often evolving scope of energy transition projects. In the traditional project management paradigm, scope changes are often viewed as challenges to be minimized due to their potential to cause delays and cost overruns. However, in the agile framework, scope flexibility is considered an asset that allows projects to evolve based on emerging insights, technological breakthroughs, or shifts in stakeholder expectations.

Agile project delivery breaks down the project into smaller, manageable segments or iterations, each with its own set of goals and deliverables. This structure allows project teams to focus on achieving specific objectives within each iteration, facilitating more detailed planning, execution, and assessment. Iterative cycles also enable the incorporation of lessons learned from previous iterations, enhancing the overall quality and effectiveness of the project.

Through iterative cycles, project teams can also identify and address risks early in the project lifecycle. Each iteration serves as an opportunity to assess potential challenges and implement mitigation strategies, reducing the likelihood of significant issues arising later in the project. This proactive approach to risk management is particularly beneficial in the energy sector, where projects often involve new technologies or are subject to complex regulatory environments.

In this section, the core aspects of agile project delivery are explored, emphasizing its advantages and some of its drawbacks in the context of the energy transition.

- Incorporating emerging technologies

The rapid pace of technological innovation in the energy sector, from advancements in renewable energy sources like wind and solar to break-throughs in energy storage and smart grid technologies, necessitates a project management approach that can swiftly incorporate new technologies. Agile project delivery frameworks are designed to accommodate such evolutions, enabling project teams to test, iterate, and scale new technologies as they become available. This iterative process allows for the exploration of various technological pathways, ensuring that the final project outcome leverages the most effective and efficient solutions available.

- Responding to regulatory and market changes

The energy sector is heavily influenced by regulatory policies and market dynamics, which can shift significantly over the course of a project. Agile project delivery allows for a more responsive approach to these changes, enabling project teams to adjust their strategies and objectives as needed. This adaptability is essential for ensuring compliance with new regulations, capitalizing on market opportunities, and mitigating risks associated with policy and market volatility.

- Stakeholder engagement and feedback loops

Agile methodologies emphasize the importance of continuous stakeholder engagement and feedback. Energy companies can better understand and address their concerns and expectations by involving stakeholders, including regulatory bodies, community representatives, and end-users throughout the project lifecycle. Regular feedback loops allow for the refinement of project objectives and deliverables, ensuring that the final outcome aligns with stakeholder needs and contributes to broader sustainability and energy transition goals.

- Continuous improvement and iterative development

Continuous improvement and iterative development are about embrac-ing a cyclical process of planning, executing, reviewing, and adjusting. This approach contrasts with the linear trajectory of traditional project management, offering several advantages in the context of energy pro-jects. In an Agile or adaptive framework, teams continuously gather feedback and make adjustments throughout the project's lifecycle, rather than waiting until the end to assess performance. This allows for flexi-bility in responding to unexpected changes, such as fluctuating market conditions, regulatory updates, or technological advancements. In the context of energy projects, which often deal with complex and rapidly changing environments, Agile methodologies offer significant benefits. For example, the cyclical review process ensures that risks are identified early and addressed proactively, minimizing costly delays or missteps. This dynamic approach fosters a more resilient project delivery framework, better suited to the unpredictable nature of energy transition projects, where the scope is not fully defined or evolving new technologies are involved.

• Leveraging data and analytics for decision-making

The use of data analytics through the deployment of data-driven approaches as outlined in Chapter 5, and digital tools are integral to the agile project delivery framework. These technologies provide project teams with real-time insights into project performance, enabling data-driven decision-making. For instance, predictive analytics can forecast potential delays or cost overruns, allowing teams to adjust their plans proactively. Similarly, digital collaboration tools facilitate seamless communication and coordination among project stakeholders, enhancing team efficiency and project outcomes.

While agile project management, a tried and tested methodology, especially suited to scenarios such as the energy transition with emerging technology and complex project scopes is celebrated for its flexibility and responsiveness, it also has its challenges.

One of its drawbacks is the perceived lack of comprehensive documentation. Agile's iterative nature, with its focus on adaptability and continuous development, often leads to minimal documentation efforts. This can create gaps in project continuity and knowledge transfer, making it difficult for stakeholders to track progress and changes comprehensively. The PMI, with its emphasis on thorough documentation and standardized processes, may find this aspect of Agile challenging to reconcile with its principles of project management.

Furthermore, Agile's inherent flexibility and scope for continuous change can sometimes result in scope creep, where projects expand beyond their original objectives due to ongoing changes and additions. This can be particularly problematic in environments that require strict project scopes and budgets. Additionally, the Agile methodology demands significant time investment and close collaboration among team members, which can strain resources and affect the availability of personnel for other projects. The requirement for continuous stakeholder engagement and the potential for unpredictability in project outcomes also pose risks to project stability and success. Undoubtedly, there is a need for project managers to carefully balance Agile's benefits with its drawbacks.

The agile project delivery framework offers a robust and flexible approach to managing the complexities and uncertainties of energy sector projects but must be applied in a balanced way. If deployed effectively, it has the potential to support energy companies to navigate the challenges of the energy transition more effectively through emphasizing flexibility, adaptability, and continuous improvement.

5. Performance monitoring and control

Continuous monitoring of project performance against established metrics and milestones enables early detection of deviations from the plan. This element involves tracking progress, managing changes, and adjusting strategies as necessary to keep the project on course. Performance Monitoring and Control is a critical facet of a Project Delivery program, especially in the management of complex projects where the scope, schedule, and resource allocation may not be fixed. This element ensures that projects adhere to their defined objectives and can respond dynamically to any issues that may threaten their success. Through the establishment of clear metrics and milestones, project managers can gauge project performance in real-time, facilitating a proactive approach to project management that emphasizes agility and informed decision-making.

The foundation of effective performance monitoring is the definition of specific, measurable, achievable, relevant, and time-bound (SMART) metrics and milestones that reflect the critical success factors of the project. These benchmarks serve as a roadmap, guiding the project through its lifecycle and providing a basis for evaluation. Metrics might include parameters, such as budget adherence, quality of deliverables, timeline progress, and stakeholder satisfaction, among others. Milestones, on the other hand, mark significant points in the project timeline, offering opportunities to assess progress, celebrate achievements, and realign priorities as necessary.

Implementing systems for real-time tracking and reporting of project performance against these metrics and milestones is essential. This may involve the use of project management software tools that provide dashboards and analytics, enabling project managers and teams to visualize progress, identify trends, and spot potential issues before they escalate. Regular reporting to stakeholders is also crucial, ensuring transparency and maintaining trust by keeping all parties informed about the project's status.

In the fluid environment of complex projects, change is inevitable. Performance monitoring and control mechanisms must, therefore, include processes for managing these changes effectively. This involves assessing the impact of potential changes on project scope, timeline, and resources, and deciding whether to integrate them into the project plan. A structured change management process helps in documenting, approving, and implementing changes systematically, ensuring that the project remains aligned with its objectives while accommodating necessary adjustments.

The real value of performance monitoring and control lies in its ability to inform strategic decision-making. By continuously assessing project performance, project managers can identify areas where the project is not meeting its objectives and initiate corrective actions. This may involve reallocating resources, revising timelines, or modifying project strategies to address underperformance. The goal is to ensure that the project remains on track to achieve its goals, even as conditions and requirements evolve.

Finally, performance monitoring and control contribute to organizational learning by highlighting lessons learned during the project. This feedback loop allows project teams and the wider organization to understand what worked well and what did not, informing future projects and contributing to a culture of continuous improvement.

In summary, Performance Monitoring and Control is a dynamic and ongoing process that plays a vital role in ensuring project success. It is crucial to establish clear metrics and milestones, and progress tracking in real-time, which enables proactive management of change and adjustment of strategies as necessary.

6. Alignment with strategic goals

Ensuring that the project aligns with the organization's strategic goals and objectives is essential in a Project Delivery Assurance program. This alignment ensures that the project contributes to the broader strategic vision, enhancing its value and relevance to the organization. This element ensures that every project undertaken is not just a standalone venture but a strategic endeavor that contributes to the broader objectives of the organization and, in the context of the energy sector, supports the transition toward sustainable energy sources.

The process begins with the integration of the project into the organization's strategic planning. This means that even before a project is initiated, its objectives are evaluated for their contribution to the strategic goals of the organization. Projects are selected and prioritized based on their potential to drive the organization forward, fulfill its mission, and achieve its vision for the future. This strategic integration ensures that resources are allocated to projects that are of strategic importance, maximizing the return on investment and ensuring that the organization's efforts are concentrated on areas of strategic relevance.

Throughout the project lifecycle, maintaining alignment with strategic goals requires continuous monitoring and adaptation. This involves regular reviews of the project's progress against its strategic objectives and the organization's changing strategic landscape. As organizations operate in dynamic environments, strategic goals may evolve, necessitating adjustments to project objectives and deliverables to maintain alignment. This adaptive approach allows organizations to respond to new opportunities, challenges, and market conditions, ensuring that projects remain relevant and aligned with the strategic direction.

Engaging stakeholders in discussions about the project's alignment with strategic goals is also important. This includes not just the project team and senior management but also external stakeholders who may influence or be affected by the project's outcomes. By fostering open communication and collaboration, organizations can ensure that diverse perspectives are considered in aligning projects with strategic objectives, enhancing buy-in, and ensuring that the project delivers value not just to the organization but to all stakeholders involved.

Lastly, establishing metrics and reporting mechanisms to measure the project's contribution to strategic goals is essential. This involves defining KPIs that reflect the project's success in achieving strategic objectives. Regular reporting on these KPIs to senior management and stakeholders not only provides transparency but also facilitates informed decision-making, allowing for course corrections as needed to enhance strategic alignment.

In essence, the alignment with strategic goals within Project Delivery Assurance ensures that complex projects are not merely executed efficiently in terms of scope, time, and budget, but they also significantly contribute to the organization's strategic objectives and long-term success. This alignment is critical for ensuring that projects deliver meaningful, sustainable value, driving progress toward strategic milestones and reinforcing the strategic direction of the company in the face of industry challenges and opportunities.

7. Quality Assurance

Quality Assurance (QA) is a key component of a Project Delivery Assurance program, playing a key role in ensuring that projects meet or exceed the established quality standards. This systematic process of verifying that project activities and outputs align with the defined requirements and specifications is essential for achieving high-quality outcomes. In this context, QA encompasses a broad range of activities, from the initial planning stages through to the final delivery, embedding a quality-oriented mindset throughout the project lifecycle. The foundation of effective QA is the establishment of clear, measurable quality standards and criteria against which the project's outputs will be evaluated. These standards should be aligned with customer expectations, regulatory requirements, and industry best practices. It is important to define these standards upfront so that project teams have a clear understanding of the quality objectives they need to achieve, guiding their efforts and decisions throughout the project.

To ensure that the project remains on track to meet these quality standards, regular reviews and audits are conducted at various stages of the project. These reviews can be internal, conducted by the project team or the organization's QA department, or external, involving third-party auditors. They provide an opportunity to assess the adherence to quality plans, identify any deviations from the quality standards, and initiate corrective actions. Regular reviews encourage a proactive approach to quality management, allowing issues to be addressed promptly before they escalate.

A key aspect of QA is the rigorous testing and validation of project deliverables. This may involve functional testing to ensure that the outputs perform as expected, as well as nonfunctional testing to assess qualities, such as usability, security, and performance. Validation of activities ensures that the deliverables meet the specified requirements, and also fulfill the intended purpose and user needs. Through comprehensive testing and validation, projects can achieve high-quality outcomes that satisfy stakeholders and contribute to the project's success.

QA is not just about meeting current quality standards but also about continuously improving quality processes and outcomes. Feedback from testing, reviews, and stakeholder inputs are analyzed to identify areas for improvement. Lessons learned are integrated into future projects, contributing to a culture of continuous improvement. Consistently enhancing QA processes allows organizations to elevate the quality of their project outputs over time, leading to better performance, increased stakeholder satisfaction, and a competitive advantage.

8. Knowledge management

Capturing and sharing knowledge gained during the project is an important part of a Project Delivery Assurance program. This includes technical libraries, lessons learned databases, best practices, and innovative solutions among others that can inform future projects.

Knowledge management within the framework of Project Delivery Assurance is a strategic process aimed at capturing, organizing, and sharing the wealth of knowledge and insights generated throughout the lifecycle of a project. This systematic approach to knowledge capture and dissemination enhances the effectiveness of current projects and serves as a valuable resource for improving the management and execution of future initiatives.

One of the critical components of knowledge management is the systematic capture of lessons learned during the project. This involves documenting both successes and failures, along with the context in which they occurred, the outcomes, and the insights gained. Regular debriefing sessions with project teams and stakeholders offer a structured opportunity to reflect on what worked well, what didn't, and why. This process helps to identify patterns and root causes of issues, enabling the organization to apply these insights into future projects.

From the accumulation of lessons learned, organizations can distill best practices that guide project management methodologies, decision-making processes, and execution strategies. Best practices are proven approaches or techniques that have led to desirable outcomes and are recommended for use in similar contexts. Organizations can ensure that valuable knowledge is accessible and actionable for all relevant stakeholders by codifying these practices into standard operating procedures, checklists, or guidelines.

Beyond lessons learned and best practices, knowledge management also plays a crucial role in identifying and disseminating innovative solutions developed during projects. These innovative solutions could be new processes, technologies, or approaches that significantly enhance project outcomes. Capturing these solutions drives organizations to encourage creative problem-solving and drive a culture of innovation, where teams are motivated to explore new ideas and approaches.

Effective knowledge management requires capturing and sharing knowledge in an accessible manner. This can be facilitated through knowledge management systems, intranets, or collaborative platforms where information is categorized, stored, and made easily searchable for future reference. Regular knowledge-sharing sessions, workshops, and training programs can also help disseminate insights and best practices across the organization, enhancing collective expertise and competency.

Ultimately, the goal of knowledge management is to enable continuous improvement across projects and the organization as a whole. The systematic capture and leverage of knowledge gained from each project enables organizations to refine their project management methodologies, enhance team capabilities, and improve project outcomes over time. This approach to learning and improvement supports organizations to remain agile, responsive, and competitive in an ever-changing project landscape.

These elements of the Project Delivery Assurance program, when effectively applied, can provide a robust framework for navigating the complexities of large-scale projects, ensuring that they are delivered successfully, meet their intended objectives, and contribute positively to the strategic goals of the organization.

Integrating Delivery Assurance throughout the project lifecycle is a strategic approach that ensures the principles of Delivery Assurance are embedded in every phase of a project, from its initiation and planning stages through to execution, monitoring, and closure. This integration allows for a continuous flow of assurance activities, ensuring that every stage of the project benefits from real-time insights and learnings. Such an approach is instrumental in identifying and addressing challenges promptly, as well as seizing opportunities that may arise during the project's journey. It facilitates a dynamic and responsive project management environment where adjustments can be made efficiently and effectively, thereby enhancing the project's ability to meet its objectives.

Risk management and quality control are pivotal elements within the Delivery Assurance framework, focusing on the early identification, thorough assessment, and strategic mitigation of risks that could potentially derail project outcomes. Employing sophisticated risk analysis techniques, project managers are equipped to foresee potential challenges and devise preemptive strategies to mitigate them. Quality control measures, on the other hand, are implemented to uphold the integrity and performance of the project, ensuring that every deliverable meets the established standards of quality. These proactive steps are crucial in the energy sector, where projects often face a myriad of risks stemming from technological advancements, regulatory changes, and market volatility.

Furthermore, the alignment of projects with the strategic goals of the organization and the broader objectives of the energy transition is a fundamental aspect of Delivery Assurance. This strategic alignment is critical for ensuring that projects not only achieve their immediate objectives but also contribute to the long-term goals of sustainability, innovation, and competitiveness within the energy sector. By integrating Delivery Assurance frameworks that prioritize this alignment, organizations can ensure that their projects are not just successful in isolation but are also meaningful contributors to the strategic direction of the company and the industry's shift toward cleaner and more sustainable energy sources. This alignment reinforces the importance of viewing projects not as standalone endeavors but as integral components of a larger strategic vision, driving progress, and innovation in the energy transition.

Project startup

I often bring up an analogy of a golfer hitting a ball off the tee when talking about project startup as I feel it is particularly appropriate to illustrate the impact of a poor project startup. At the tee, a golfer takes great care to ensure that everything from stance to swing is aligned, knowing that even a minor five-degree error can send the ball far off course, requiring significant corrective measures. This deviation will only be fully realized when the ball is 200 yards down the fairway in the rough! Similarly, a project startup must be precise, well-considered, and calculated to set the course for success. If the startup phase is off by even a small margin, the project can end up in a hugely different position than intended, necessitating costly and time-consuming corrections.

The initiation of any project is a complex blend of planning, mobilization, and strategic alignment, with the project startup phase taking center stage. This stage is the bridge between the conceptualization of a project and its materialization into action, serving as the launching pad from which all subsequent project activities are propelled. During this critical stage, the foundations laid will bear the weight of all future endeavors, and any missteps here can ripple throughout the life of the project, often with magnified consequences. It is a time when the project's vision starts taking a tangible form through the establishment of governance structures, the development of detailed plans, and the alignment of resources to project objectives.

Recognizing the importance of this stage, project startup also emerges as a cornerstone of any Project Delivery Assurance program. It is in this stage that the theoretical becomes practical, where the detailed project plan, risk management strategies, and stakeholder commitments are tested against the harsh light of reality. It is the period where initial assumptions are validated, project teams are assembled, and the mechanisms of project control are calibrated. Ensuring a robust startup phase is pivotal to the successful commencement and seamless execution of a project. It sets the tempo for how the project will unfold and establishes the protocols that will guide its progression.

The significance of the project startup stage is highlighted by numerous studies, including those conducted by the PMI, which emphasize its role in the overall success of projects. Time and again, research into project delays and failures reveals that many of the challenges leading to unsuccessful project outcomes can be traced back to this phase. A failure to accurately establish clear objectives, an underestimation of the resources required, and inadequate risk assessment during project startup are frequently cited reasons for why projects falter. Therefore, paying diligent attention to this phase is not just recommended practice but a critical strategy in safeguarding against the pitfalls that have historically led to project overruns, scope creep, and outright project failure.

An effective startup approach

An effective project startup approach should be characterized by meticulous planning and the integration of digital tools. These tools facilitate communication, planning, tracking, and reporting, allowing for a more streamlined and efficient startup process. Digital dashboards, for instance, can provide real-time visibility into the startup phase, enabling timely decisions and interventions.

Additionally, the startup phase must be a documented procedure. This documentation serves as a blueprint for the project and ensures consistency across various projects within an organization. Documenting the startup process allows for a structured approach that can be followed and refined over time, incorporating best practices and lessons learned from previous projects.

Key components of a project startup

Having a standardized procedure documented ensures consistency in the startup process. A startup checklist is an essential tool that guides project managers and teams through the necessary steps, ensuring that no critical component is overlooked.

Utilizing digital tools provides a platform for collaboration, task management, and data analysis. A startup dashboard offers a visual representation of the project's status and helps track important milestones and deliverables within the first 90 days, which is often considered the critical period for project startup.

A robust startup approach includes a mechanism to capture and incorporate lessons learned into the project startup process. This feedback loop is invaluable for continuous improvement and helps avoid repeating past mistakes.

Engagement from all levels of the organization and in particular functional leaders and Subject Matter Experts is key. The startup phase requires input and commitment from various stakeholders, ensuring alignment and buy-in from the outset. Recognizing that project teams may not be fully mobilized at the time of award is important, it is advisable to have functional overhead, such as operations and PMO support, to bridge the gap and support project startup activities. This additional support can provide the necessary resources and expertise to ensure that startup activities are completed successfully.

The project startup is not just another stage in the project lifecycle and it is the stage that can determine the overall success of the project. Ensuring that it is executed with precision, supported by digital tools, and well-documented, with full engagement from the organization, is not just beneficial, it is essential. Like the golfer who knows the value of the tee shot, project managers must appreciate the significance of a well-executed project startup. It is the difference between a project that is off target and one that is on the path to success from day one.

Delivery Assurance reviews

Project Delivery Assurance reviews are systematic evaluations conducted at various stages of a complex project to ensure it remains on track to meet its objectives, adheres to quality standards, and aligns with organizational goals. These reviews are crucial for identifying risks, addressing issues, and implementing corrective actions in a timely manner. Typically, these reviews can be categorized into different levels, each serving a specific purpose and managed through distinct processes.

The accompanying project lifecycle graph delineates the relationship between cost and staffing levels, risk and uncertainty, and the cost of changes over the course of a typical project's lifecycle. The curves illustrate the dynamic nature of project management and highlight the importance of early (and comprehensive) planning and risk management. It highlights that changes are less costly when identified early and that risk and uncertainty can be mitigated through proper project organization and execution. The closing phase serves as a warning to avoid late-stage changes that can significantly impact the project's budget and timeline (Fig. 7.1).

Cost and staffing level curve

This curve typically follows a "bell-shaped" or "mountain-shaped" pattern. It starts low, increases as the project ramps up, peaks during the main execution phase, and finally decreases as the project winds down. The initial low

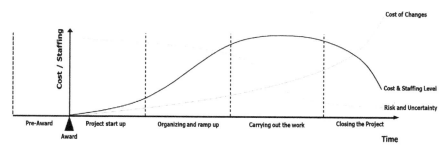

Fig. 7.1 Project lifecycle.

reflects minimal activity before the project officially kicks off. Once the project starts, costs and staffing levels rise as team members are hired, resources are procured, and activities commence. The peak occurs during the height of project activity when most of the resources are deployed, and work is being carried out at its maximum pace. As the project nears completion, staffing levels and associated costs decline as fewer people are needed, and tasks are being finalized.

Risk and uncertainty curve

This curve typically has an inverse relationship with the cost and staffing level curve. It starts high because, at the beginning of a project, there are many unknowns. The risk is inherent in questions like "Will the project proceed as planned?" "Are the estimates accurate?" and "Have all contingencies been accounted for?" As the project progresses through its planning and execution phases, these questions are answered, and the risk is managed and mitigated, so the curve decreases. Best practices in project management involve early and thorough risk assessments, which help in identifying potential problems before they occur, allowing the team to mitigate risks proactively. The risk is lowest toward the end of the project, where most uncertainties have been addressed, and the outcomes are highly predictable.

Cost of changes curve

This curve starts low and increases sharply toward the end of the project. This represents the concept that changes are relatively easy and low-cost to make early in a project when plans are still flexible and nothing has been built yet. However, as the project progresses, systems and structures get locked in, dependencies are established, and the project's direction becomes more solidified. At this point, making changes can be very costly and disruptive. In the final stages of the project, changes are most expensive because they can necessitate rework, may impact completed parts of the project, and can cause delays. This is why project managers emphasize the importance of getting requirements right early on and why agile methodologies encourage early and frequent testing to catch issues before they become costly to fix.

The relationship between these three curves is crucial for project managers to understand. For instance, they need to know that increasing staffing levels may decrease risk but might not reduce the cost of changes if the changes occur late in the project. Similarly, a high level of uncertainty in the early phases may require a more significant investment in risk management activities, which could increase costs upfront but potentially save money in the long run by avoiding expensive late-stage changes.

The lifecycle is broken down into five key stages: Pre-Award, Award, Project Startup, Organizing and Ramp-Up, Carrying out the Work, and Closing the Project, which are explained next. In terms of the application of Delivery Assurance, each stage considers targeted Delivery Assurance processes and reviews relative to that stage in order to drive Project Delivery Assurance. Delivery Assurance reviews are proposed next but clearly need to be tailored to company and project-specific requirements and are non-exhaustive. There are important reviews that should also be implemented across all stages within a live project, which are outlined at the end of this section.

1. Pre-Award

This is the initial phase before the project has been officially given the go-ahead. It encompasses all preparatory activities, such as feasibility studies, preliminary project planning, proposal writing, and budget forecasting. At this stage, project costs and staffing levels are typically at their lowest because the project has not yet started; however, the risk and uncertainty are relatively high, as the project's future is not yet secured, and numerous assumptions about scope, timelines, and costs are yet to be tested and confirmed.

In terms of risks, this phase may contend with uncertainties surrounding project funding, stakeholder buy-in, regulatory approvals, and market conditions, among others. It is a phase that requires flexibility and a forward-thinking approach, as decisions made here can significantly shape the trajectory of the entire project.

Before a project is officially awarded, it is important that a Feasibility Review is conducted, to verify that the concept is viable. A feasibility review evaluates whether the proposed solutions are practical and whether the project can deliver the intended value within the constraints of budget and time. This review should assess technical, economic, legal, operational, and scheduling aspects of the project.

A stakeholder analysis review maps out all potential stakeholders, their interests, and their influence on the project. It ensures that the project has considered the needs and expectations of each stakeholder group, preparing a plan for engagement and communication throughout the project lifecycle.

An initial scope review is key to define the boundaries of the project from the start. This review examines the preliminary scope defined for the project, ensuring that it is clear, focused, and manageable. It should involve evaluating the scope against the available resources and the project objectives to ensure alignment.

A resource capability review looks at the human and material resources anticipated for the project. It assesses the organization's capacity to supply these resources or identifies needs for external procurement. Ensuring that the project has access to the necessary skills and resources to complete the work is crucial for its success.

2. Award

Project award is a critical milestone in the project lifecycle, marking the transition from proposal to actual project kickoff. This is when the project receives official approval and the necessary funding to move forward. It signifies a commitment from sponsors and stakeholders and usually triggers contractual agreements and the formalization of project plans. While the risks and uncertainties of whether the project will proceed are largely resolved by this stage, new risks and uncertainties associated with execution come into focus. The award phase lays the foundation for the project's future, making it essential to review the project's alignment with strategic goals and its readiness for the next stages.

Once the project is awarded, it is important to ensure that all contracts reflect the agreed-upon scope, deliverables, timelines, and terms and conditions. This review involves a meticulous analysis of contract documents to verify that they are correctly drafted and that they align with the project's business case, budget, and resource plans. It also assesses whether the project team understands the contractual obligations, milestones, and performance criteria.

The success of a project often hinges on the support and engagement of its stakeholders, and therefore, a review of stakeholder commitment to the project and the effectiveness of the communication plan is advisable. This ensures that the roles and expectations of all parties are clearly defined and understood, and that there is a robust plan in place for managing stake-holder communication throughout the project lifecycle.

3. Project startup

The project startup phase signifies the transition from planning to action. During this phase, the project's baseline plans are developed in detail, resources are fully mobilized, the project team is assembled, and initial project activities commence. This stage is critical as it sets the operational tempo and establishes the frameworks within which the project will progress. Cost and staffing levels begin to rise in this phase, as the team engages in detailed planning, procurement, and establishment of the project infrastructure. Concurrently, risk and uncertainty start to diminish as the project's plans take a more definitive shape and the project's foundational elements are put into place.

The Startup phase lays the groundwork for all future efforts and is instrumental in building a strong project structure. Thus, any deficiencies in planning, team capabilities, or resource allocation can have lasting impacts on the project's trajectory.

Once the project has been awarded, it is important to ensure that the project plan is robust and comprehensive. This review involves validating the project's scope, objectives, schedule, cost estimates, and resourcing plans against the project's mandate and the organization's capabilities. It checks for coherence between various planning elements and confirms that the project plan is feasible, with a clear roadmap for execution.

As the project gears up, it is essential to verify that the team is properly staffed with the required skills and that resources are appropriately allocated. This review examines whether the staffing plan meets the project's needs, assesses the availability and readiness of critical resources, and confirms that the onboarding process for new team members is well-defined and effective. It also looks into the establishment of project offices, procurement of necessary equipment and materials, and the setup of essential systems and tools for project management.

While a preliminary risk assessment is conducted in the Pre-Award stage, the Startup stage demands a reevaluation of risks and issues based on the detailed project plans and the current state of the project. This review scrutinizes the risk management plan for adequacy and operational readiness, evaluates the processes put in place for ongoing risk identification and monitoring, and examines the initial issue log for comprehensive coverage of potential challenges. It also checks that there are clear escalation paths and resolution strategies in place.

Startup stage Delivery Assurance reviews are crucial for ensuring that the project starts on a solid footing, with a validated plan, a ready and capable team, and an established process for managing risks and issues. They serve to reinforce the project's alignment with strategic objectives and to provide confidence that the project is ready to proceed into the execution phases with a strong likelihood of success.

4. Organizing and ramp-up

The Organizing and Ramp-Up stage is a period of intensive preparation and scaling up as the project moves from initial planning and setup into full-scale execution. During this phase, the project's management processes and teams are finalized, and the project's infrastructure, both physical and procedural, is established or expanded to accommodate the upcoming increase in activity.

This is when the project's governance structures are tested and optimized, communication channels are established, project control systems are implemented, and the complete project team is brought up to speed. Costs and staffing levels continue to climb as the project gains momentum, while risk and uncertainty should be on a steady decline as the project becomes more concrete and structured.

As processes and systems are set up, it is key to ensure they integrate seamlessly to support project execution. This review focuses on evaluating the project's management processes, including communication plans, QA protocols, and control systems, to confirm that they are fully functional and effectively integrated. It assesses the readiness of the project management office (PMO) and whether project tracking tools are in place and properly configured to provide accurate reporting and facilitate decision-making.

A Team Integration and Performance Review may be conducted to assess how well the project team is coming together, examine the team structure, leadership roles, and the integration of various specialist groups within the project. It ensures that all team members understand the project objectives, their individual roles, and how they contribute to the project's success. It also looks at team performance metrics and evaluates whether the team is working cohesively to achieve project milestones, with a focus on interteam communication and problem-solving effectiveness.

As the project structure solidifies, it is advisable to reaffirm that the governance framework is robust and that the project remains compliant with all relevant regulations and standards. This review involves an audit of governance structures, a check for alignment with best practices, and a verification that the project is adhering to industry standards, legal requirements, and ethical norms. It should also ensure that the project has established clear escalation paths for issues that may arise, along with checks for the effectiveness of decision-making processes within the project's governance framework.

By conducting these reviews, the organization can be assured that as the project transitions into a period of higher activity, it is well-organized, equipped with effective systems, guided by a coherent team, and governed by a strong compliance and decision-making structure. These reviews are key to smoothing out any operational wrinkles and ensuring that the project is primed for the more demanding work that lies ahead in the execution phase.

5. Carrying out the Work

The "Carrying out the Work" stage is where the project's plans and strategies are put into action. This is the execution phase, characterized by active project work to deliver the intended outcomes. During this time, the project's costs and staffing levels typically reach their peak as all resources are engaged, and tasks are being performed at full scale. While risk and uncertainty should be at a much lower level at this stage due to the clarity that comes with ongoing operations, careful monitoring and control are still essential to navigate any deviations from the plan and to manage any issues that arise promptly.

Regular Performance and Progress Reviews focus on evaluating the progress of the project against its schedule, budget, and scope. It is a continuous process that involves tracking milestones, financial expenditures, and the quality of deliverables. This review helps to ensure that the project remains on track, identifies any variances from the plan, and triggers the necessary corrective actions if there are any discrepancies.

Regular QA Reviews are critical in ensuring that the project's outputs meet the required standards and are fit for purpose. This involves an audit of the project's processes and deliverables against predefined quality criteria. It includes reviewing testing protocols, change management procedures, and quality control measures. The aim is to identify any areas of noncompliance or potential quality issues before they become significant problems.

It is important to conduct these reviews regularly to ensure the project team maintains a tight grip on the project's pulse, ensuring that performance is in line with the project's objectives and that the quality of the output is not compromised. These reviews are essential for delivering a project successfully, on time, within budget, and to the expected quality standards.

6. Closing the project

In the final phase, the project is being closed out. Costs and staffing levels begin to fall as less work is required and the project begins to shut down operations. However, the cost of changes curve increases sharply, indicating that making changes during this late stage of the project is significantly more expensive due to the disruptions and alterations they would cause to the nearly completed work.

A project close out review can be conducted at the project's conclusion. The close out review ensures that all project objectives have been met, deliverables have been handed over, and any remaining issues have been addressed. It also focuses on capturing lessons learned and recommendations for future projects. This review is managed by preparing a comprehensive project closure report, including performance analysis, lessons learned, and final project metrics, which is then presented to stakeholders and senior management.

Each level of Project Delivery Assurance Review involves meticulous planning, documentation, and stakeholder engagement to ensure comprehensive evaluation and management. These reviews are integral to maintaining control over complex projects, ensuring alignment with organizational strategies, and ultimately achieving successful project outcomes (Table 7.1).

Table 7.1 Assurance reviews.

Assurance review	Description and application
Initiation review	Conducted at the project's outset to validate the business case, objectives, and project plan. Ensures alignment with strategic goals and evaluates the initial risk assessment. Managed through detailed documentation and approval by a steering committee.
Startup review	Once the project has been awarded, it is important to ensure that the project plan is robust and comprehensive and conduct a Startup review. This review involves validating the project's scope, objectives, schedule, cost estimates, and resourcing plans against the project's mandate and the organization's capabilities. It checks for coherence between various planning elements and confirms that the project plan is feasible, with a clear roadmap for execution.
Design review	Takes place after the design phase is complete. It evaluates design choices, technical solutions, and the implementation approach to ensure feasibility and alignment with project objectives. Managed through technical reviews and stakeholder consultations.

Continued

Table 7.1 Assurance reviews—cont'd

Assurance review	Description and application
Risk assessment review	Conducted at an early stage in the project identify and manage potential risks. A detailed risk assessment is conducted, examining every aspect of the project from resources to regulatory concerns. The purpose is to proactively develop mitigation strategies for identified risks and to establish a risk management plan.
Gate reviews (stage-gate reviews)	Critical checkpoints before moving to the next phase, ensuring deliverables are complete and the project is ready to proceed. Managed by reviewing completed work against phase objectives and deciding on the project's progression.
Close out review	Conducted at project completion to ensure all objectives have been achieved, and deliverables handed over. Focuses on capturing lessons learned. Managed through the preparation of a closure report presented to stakeholders.
Monthly project reviews	Regular, recurring reviews to assess ongoing project performance, identify any variances from the plan, and adjust forecasts and strategies accordingly. Managed through monthly status reporting and meetings with project teams and key stakeholders.
Peer reviews (based on criticality)	Involves the evaluation of project work by peers, especially for critical project components. Helps in identifying potential issues, sharing knowledge, and ensuring quality standards. Managed through structured sessions where project materials are examined and feedback is provided.

The role of collaborative contracting in Delivery Assurance

In the face of the energy sector's complexities and the ambitious goals of the energy transition, traditional contracting methods are increasingly seen as insufficient. Collaborative contracting as presented in Chapter 3, which emphasizes mutual goals, shared risks and rewards, and partnership over simple transactional relationships, is gaining traction. This approach is particularly suited to the sector's dynamic nature, where projects often involve cutting-edge technologies, require significant capital investment, and are subject to strict regulatory environments. Collaborative contracting can significantly enhance project delivery by fostering a more integrated, transparent, and cooperative framework among all project stakeholders.

The shift from transactional to partnership approaches in contracting signifies a fundamental change in how projects are conceived, developed, and executed. Unlike traditional contracts, which often focus on delineating scopes, timelines, and penalties in a rigid manner, collaborative contracts seek to build relationships where all parties are invested in the project's success.

Central to collaborative contracting is the development of trust and a shared vision among project participants. This involves open communication, transparency, and a commitment to mutual success. Collaborative contracts can help ensure that projects are completed successfully and also deliver greater value to all stakeholders, including the community and the environment by focusing on aligning the goals of all parties. This includes mechanisms for shared risk management, where risks are identified, assessed, and allocated in a manner that is fair and proportionate to each party's ability to manage them. This approach encourages joint problem-solving and innovation, as parties are incentivized to work together to mitigate risks and overcome challenges, rather than shifting blame or responsibility.

Projects utilizing collaborative contracts frequently experience fewer delays, lower costs, and higher-quality outcomes. This is attributed to the collective focus on efficiency and excellence, with parties more willing to share resources, knowledge, and expertise to achieve the best possible results. One of the key reasons for this is that collaborative contracts by nature are cooperative and foster an environment where innovation is encouraged and rewarded. Parties are more likely to explore new technologies, processes, and solutions when they know that the benefits, as well as the risks, will be shared. This is particularly advantageous in the energy sector, where technological advancements can significantly impact project success and sustainability.

Collaborative contracting represents a paradigm shift in the execution of projects aligning with the sector's move toward more sustainable, efficient, and innovative practices. Partnership, shared objectives, and mutual respect among stakeholders enable collaborative contracts to significantly contribute to the successful delivery of complex energy projects. Coupled with the strategic use of digital and data-driven tools, this approach can ensure that projects are completed on time and within budget and also contribute positively to the energy transition and the broader goals of sustainability and social responsibility.

Leveraging digital and data-driven approaches

Chapter 5 introduced the concept of a data-driven organization, emphasizing the importance of leveraging data to innovate, optimize, and lead in today's rapidly evolving energy sector. With advancements in digital solutions, low code applications, and computing power, organizations are grappling with unprecedented amounts of unstructured or semistructured data from various sources from financial logs, text files, multimedia forms, sensors, and instruments. Simple data visualization tools are insufficient for processing such vast volumes and varieties of data, necessitating the use of more sophisticated analytical tools and algorithms. As organizations within the energy sector strive to become more data-driven, the significance of effective Delivery Assurance programs becomes increasingly apparent. These programs are essential for ensuring the successful and efficient delivery of capital projects amidst the myriad challenges and opportunities present in the evolving energy landscape.

Integrating data-driven approaches into Delivery Assurance processes enables organizations to enhance project management, mitigate risks, and optimize resource allocation. Furthermore, leveraging data analytics enables proactive identification of potential issues, facilitates real-time decision-making, and enhances overall project performance. The synergy between data-driven organizations and Delivery Assurance programs is clear and paramount for achieving operational excellence and driving sustainable growth.

Integrating digital tools with a Delivery Assurance Program

The energy sector's journey through the transition toward more sustainable and efficient energy systems is increasingly underpinned by digital and data-driven technologies. These technologies are transforming project delivery, offering new ways to plan, execute, monitor, and evaluate energy projects with unprecedented precision and insight. Leveraging digital tools and data analytics enables energy companies to navigate the complexities of the energy transition, optimize project outcomes, and ensure alignment with broader environmental and sustainability goals. This integration leverages the precision, speed, and analytical power of digital technologies to enhance the effectiveness of Delivery Assurance practices, ensuring that projects meet their objectives and also align with broader strategic goals and sustainability objectives of the organization.

Some examples of how digital tools can bolster Delivery Assurance, making project delivery more reliable, efficient, and adaptable to the challenges of the energy transition are presented next.

A robust and integrated Project Management (PM) system plays a pivotal role in enhancing Delivery Assurance by providing a comprehensive platform for managing various elements of project execution. These PM platforms can vary significantly in terms of complexity and features, from a basic planning software to a full-fledged system integrating document control, planning, cost management, resource allocation, communication, action tracking offering project teams a centralized hub for collaboration, project monitoring, actions tracking, and decision-making. Through real-time visibility into project progress and milestones, stakeholders can better coordinate efforts and address issues promptly. Automated alerts and workflow management features further bolster project reliability by ensuring that tasks are completed on time and deviations from the plan are swiftly identified and rectified. There are numerous PM systems available on the market today and cater for a variety of different project services organizations.

Data analytics and predictive modeling serve as indispensable tools for project teams navigating the intricacies of energy projects. These tools encompass a variety of software solutions designed to extract insights from project data and forecast future outcomes. Again there are numerous software titles commercially available, which enable project teams to visualize and analyze project data through interactive dashboards and reports. Project teams can identify trends, patterns, and outliers within their project data, facilitating informed decision-making. For example, Microsoft Power BI offers capabilities for data visualization and analysis and allows project teams to gain a deeper understanding of their project performance, identify areas for improvement, and make data-driven decisions to mitigate risks and optimize project outcomes.

In terms of predictive modeling, software is commonly used by project teams to build predictive models based on historical project data. These tools employ advanced statistical algorithms and machine learning techniques to identify relationships between project variables and forecast future outcomes. For instance, project teams can use predictive models to anticipate potential schedule delays, cost overruns, or equipment failures, allowing them to proactively address these issues before they occur. Additionally, tools, such as MATLAB and Python, provide flexible and customizable frameworks for building predictive models tailored to specific project requirements. Leveraging these predictive modeling tools allows project teams to enhance Delivery Assurance by anticipating risks, optimizing resource allocation, and ensuring project success in the face of uncertainty.

Through the analysis of project data, these tools, if deployed effectively can provide invaluable insights into patterns, trends, and potential risks, equipping teams with the knowledge needed to make informed decisions. For instance, by examining historical project performance data, teams can identify recurring issues or inefficiencies and take corrective actions to mitigate future risks. Moreover, data analytics tools enable teams to uncover hidden correlations between project variables, shedding light on potential risks that may have otherwise gone unnoticed.

Predictive modeling takes risk management a step further by utilizing current and historical project data to forecast future outcomes, leveraging advanced statistical algorithms and machine learning techniques. This enables predictive models to anticipate challenges and opportunities, enabling teams to implement preemptive measures to mitigate risks and capitalize on opportunities. For example, predictive models can forecast potential schedule delays or cost overruns, allowing project teams to adjust their plans accordingly to minimize disruptions and optimize resource allocation. In the volatile energy sector, where market conditions and regulatory landscapes are constantly evolving, this data-driven approach to risk management is crucial for maintaining project timelines and budgets, ultimately ensuring successful project delivery in an unpredictable environment.

Digital twins represent a groundbreaking advancement in Project Delivery Assurance, offering project teams the ability to create virtual replicas of physical projects that mirror their real-world counterparts. These digital twins serve as powerful simulation environments where project teams can conduct real-time analysis and testing, allowing for the exploration of various scenarios and the optimization of project designs before implementation. Through digital twins, project teams gain valuable insights into how changes in design or operations can impact project performance, enabling them to make informed decisions to mitigate risks and enhance operational efficiency. This technology proves particularly invaluable in the for complex energy projects, where the stakes are high and the margin for error is slim. Digital twins enable project teams to assess the viability of new technologies or configurations, identify potential bottlenecks, and optimize resource allocation to ensure project success (Fig. 7.2).

Moreover, digital twins offer a holistic view of project performance, empowering project teams to make data-driven decisions throughout the project lifecycle. Integrating real-time data from sensors and other sources into the digital twin environment enables project teams to continuously monitor and analyze project performance, identifying issues as they arise and proactively addressing them to maintain project timelines and budgets.

Fig. 7.2 Digital twin.

An example by Kent [2] is the concept of an "8D Digital Twin," which is a multidimensional representation of a physical asset, process, or system integrating various "dimensions" of data that complement the digital twin concept, laid out in the image next, highlighting their interconnected nature.

Starting from "1D," which includes Metadata Documents, and moving clockwise, the dimensions progress through "2D" with Drawings and Native Files, "3D" representing the core Model, "4D" that ties in Schedule, "5D" incorporating Cost, "6D" for Real Time Data, "7D" indicating Live Streaming, and culminating in "8D" with ML Predictive Data.

The digital twin concept can be deployed in this manner in a comprehensive system that integrates from basic documentation to advanced predictive analytics, facilitating a real-time, data-driven approach to asset management and monitoring (Fig. 7.3).

Digital twins facilitate collaboration among project stakeholders by providing a centralized platform for sharing information and insights, and enhance project reliability, efficiency, and adaptability by providing project teams with the tools they need to navigate the complexities of energy projects with confidence and precision.

Data-driven companies can utilize the power of data analytics to transform their approach to Delivery Assurance, ushering in a new era of proactive risk management and optimized project performance. Through sophisticated risk analysis facilitated by data analytics, project teams gain unparalleled insights into the intricacies of their projects. By harnessing

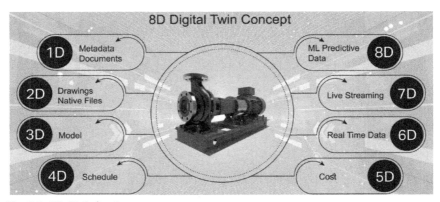

Fig. 7.3 8D Digital twin concept.

real-time data, these teams can identify, assess, and prioritize risks with precision, enabling them to allocate resources strategically and focus efforts
where they are most needed. Predictive analytics emerges as a game-changer
in this context, offering the ability to forecast potential issues before they
escalate into major setbacks. Whether it is anticipating equipment failures,
forecasting supply chain disruptions, or identifying market volatility, predictive analytics empowers project teams to stay ahead of the curve and implement proactive risk mitigation strategies. Data-driven companies can
anticipate challenges, optimize resource allocation, and ultimately enhance
the overall Delivery Assurance process by utilizing historical data, deploying
machine learning, and advanced statistical models.

The integration of Delivery Assurance with digital tools is transforming
project management in the energy sector. Energy companies can achieve
greater efficiency, transparency, and adaptability in their projects by
harnessing the capabilities of advanced software, data analytics, and digital
twins. This digital approach has been proven to improve project delivery
outcomes and also support the strategic objectives of the energy transition,
paving the way for a more sustainable and resilient energy future.

The energy sector stands at a pivotal juncture, with the transition toward
sustainable and renewable energy sources presenting both immense challenges and opportunities. This transition, driven by global environmental
concerns, technological innovation, and changing regulatory landscapes,
necessitates a reevaluation of traditional project management practices. As
we have explored through various facets of project delivery, including agile

methodologies, collaborative contracting, digitalization, and the imperative of Delivery Assurance, it is clear that a new paradigm in project management is emerging—one that is adaptive, collaborative, and technologically advanced.

Agile project delivery has proven to be a cornerstone in managing the complexities and uncertainties inherent in energy projects. Its emphasis on flexibility, adaptability, and responsiveness allows project teams to navigate changing project scopes, emerging technologies, and fluctuating market demands effectively. Adopting an iterative approach to project planning and execution enables energy companies to ensure that projects remain aligned with strategic objectives, even as those objectives evolve in response to new insights or external pressures.

Collaborative contracting has emerged as a key strategy for fostering partnerships that extend beyond mere transactional relationships, prioritizing mutual goals, shared risks, and rewards, and a commitment to joint problem-solving. This approach enhances project outcomes and strengthens the resilience and adaptability of project teams, making them better equipped to tackle the multifaceted challenges of the energy transition.

The integration of digital and data-driven approaches into Project Delivery Assurance represents a significant leap forward in how projects are conceptualized, executed, and monitored. Digital tools, from project management software and digital twins to data analytics and IoT technologies, provide unparalleled visibility, control, and insight into project operations. These technologies enable real-time decision-making, predictive risk management, and enhanced stakeholder engagement, driving efficiency, transparency, and sustainability in project delivery.

As the energy sector continues to evolve, the principles of agile project delivery, collaborative contracting, and digitalization will become increasingly integral to the successful realization of energy projects. These strategies offer a blueprint for navigating the energy transition, enabling energy companies to deliver projects that not only meet but exceed expectations in terms of efficiency, effectiveness, and alignment with global sustainability goals. The journey toward a sustainable energy future is fraught with challenges, but it also offers unparalleled opportunities for innovation and transformation. Embracing lessons learned from real-world applications and the insights gleaned from emerging practices permits energy companies to position themselves at the forefront of the energy transition. The future of project management in the energy sector is one of collaboration, innovation, and resilience, guided by the strategic application of agile methodologies, collaborative approaches, and cutting-edge digital technologies.

References

[1] Reuters, Reuters Article, 'Saudi's NEOM Says to Be World's First Fully Renewable Utility-Scale Wind-Solar Projects', 2022.
[2] W. Ghadban, Digital Twin—A Virtual Revolution in oil & gas, Kent PLC, 2023.

CHAPTER EIGHT

Challenging the norm

Introduction

To introduce this chapter, I'm going to borrow an analogy from Formula 1 racing to illustrate the concept of "challenging the norm." In the high-stakes environment of Formula 1, where continuous improvement is not just beneficial, it is essential for survival and success, much like the new energy industry landscape that we find ourselves in. In this context, the transformation seen in Formula 1 over the years is something to marvel at. Formula 1 pit stops for example over the decades serve as a striking metaphor for what energy companies must achieve in today's rapidly evolving landscape.

Historically, Formula 1 pit stops were leisurely affairs, with a typical stop in the 1950s lasting up to 67 s. Teams would methodically refuel the car, change tires, and make necessary adjustments, with what seemed like no pressing urgency (Fig. 8.1).

However, as the sport evolved, the competition grew fiercer, and the margins for victory became ever thinner, making every second on the track critical. This shift in the racing paradigm transformed pit stops from routine maintenance breaks into crucial, strategically significant events. Teams began to innovate aggressively, streamlining their processes and leveraging cutting-edge technology to shave off every conceivable fraction of a second. Today, pit crews can execute a full stop in under 2 s, exemplifying a dramatic enhancement in operational efficiency (Fig. 8.2).

This evolution in Formula 1 is analogous to the current challenges faced by the energy sector, particularly with the ongoing energy transition. Like Formula 1 teams, energy companies are finding that traditional methods of operation are no longer sufficient in a landscape marked by rapid technological advances and increasing environmental regulations. The push toward renewable energy sources, the integration of new technologies, and the imperative to reduce carbon footprints demands a paradigm shift in how energy projects are managed and executed.

Powering Through the Transition
https://doi.org/10.1016/B978-0-323-91754-4.00010-8

Fig. 8.1 1950s Formula 1 pit stop.

Fig. 8.2 2024 Formula 1 pit stop.

Traditional energy companies often operate within established norms that prioritize stability over speed and innovation. However, the pressing need for sustainability and efficiency in the face of climate change and technological innovation demands a reevaluation of these norms. Just as Formula 1 teams reengineered every aspect of the pit stop process through better tools, training, and teamwork strategies, energy companies must embrace a similar ethos of innovation and efficiency.

To thrive amidst these changes, energy companies must challenge the norm and continuously seek out improvements in every facet of their operations. This means optimizing not just the energy production processes but also how projects are managed, how data is utilized, and how operations are streamlined to enhance efficiency and sustainability. The goal is to turn every operational element, no matter how small, into an opportunity for strategic advantage, much like how Formula 1 teams view their pit stops.

Embracing this mindset of continuous improvement is an imperative for energy companies to survive and ultimately lead in the new era of the energy transition. This requires a cultural shift toward innovation and agility, drawing lessons from high-performance environments like Formula 1 where the cost of complacency is not just a lost race, but potential obsolescence.

The continuous improvement imperative

In the transformative landscape of the energy transition, the imperative of continuous improvement compels companies to "challenge the norm" and reimagine their operational strategies. This era is not just about adopting new technologies but fundamentally reshaping how energy services are delivered and managed. The drive toward more sustainable and efficient systems is fraught with challenges but brimming with opportunities for innovation and leadership.

Continuous improvement serves as a strategic framework within which companies can iteratively test, refine, and perfect new methodologies. This process is crucial not only for enhancing operational efficiency but also for building the agility necessary to respond to emerging technologies and shifting market dynamics effectively. Engaging continuously in improvement processes helps energy companies to enhance their competitiveness.

The concept of a Center of Excellence (CoE) as presented in Chapter 6 integrates seamlessly into this framework, acting as a catalyst for sustained improvement and innovation. As a quick recap, a CoE is a dedicated team or entity focused on promoting best practices, fostering innovation, and developing capabilities that support business goals, particularly in the context of the energy transition. This entity works as a hub of expertise, pulling together skills, knowledge, and resources across the organization to drive significant improvements in efficiency and performance.

CoEs play a pivotal role by centralizing expertise. They gather specialized knowledge and skills, becoming the epicenter for advanced learning and thought leadership within the company. This centralization helps in standardizing processes and ensures the dissemination of best practices. They drive innovation by focusing on cutting-edge technologies and methodologies, pushing the boundaries of what's possible, and driving the development and implementation of innovative solutions that keep the company at the forefront of the industry. They enhance agility by developing strategies and processes that enable the organization to swiftly adapt to changes in technology and market conditions, ensuring the company remains resilient and responsive. They build collaborative networks by extending their influence, forming alliances with academic institutions, industry groups, and other entities, fostering a collaborative ecosystem that supports sustained innovation and growth. Incorporating a CoE into the continuous improvement strategy ensures that energy companies are leaders shaping the future of the industry. Institutionalizing the pursuit of excellence and innovation enables organizations to build a culture that challenges the norm and sets new standards, thereby defining the cutting edge of the industry.

This chapter sets the stage for a detailed exploration of how challenging the norm is a critical facet of a Company's DNA and integrates with a CoE. Just as the step change advances in a Formula 1 pit stop can determine the outcome of a race, so too can streamlined and enhanced operations determine the success of energy companies in a transitioning market. Challenging the norm requires a laser focus on measuring performance, establishing and putting in place robust systems of measurement and management.

The Paradigm of Measurement and Management
The legacy of Peter Drucker

Peter Drucker's renowned saying, "what gets measured gets managed," can be regarded as a foundation for operational efficiency, as well as transformative force in organizations facing profound transformation, such as the energy sector. This principle highlights the necessity for accountability and serves as a catalyst for challenging conventional norms by compelling organizations to continuously scrutinize and enhance their operational frameworks.

Drucker's saying does more than highlight the importance of accuracy and accountability in business processes, it actively demands it. In the context of the energy sector, which is pivotal to global sustainability efforts, this means implementing a sophisticated tracking system that meticulously monitors key performance metrics. Depending on each organizations intricacies, these may include energy efficiency, carbon emissions, resource allocation, project delivery, and sustainability initiatives. Clearly defining these metrics enable organizations to have the tools to benchmark their current performance, as well as the data needed to forge a clear path for future advancements.

This commitment to measurement fosters a culture of continuous improvement within companies. Strategic, data-informed decisions begin to replace reactive guesswork, embedding a proactive approach to management across all levels of the organization, as we discussed in Chapter 5—the data-driven organization. This shift is crucial in a sector where technological advancements and regulatory requirements are continually evolving. The ability to quickly adapt to new information or changing circumstances can significantly enhance a company's operational agility and responsiveness.

The application of Drucker's principle extends beyond internal processes. It influences external perceptions and relationships as well, playing a critical role in shaping a company's reputation with stakeholders, including investors, regulatory bodies, and the general public. Transparent measurement and management practices can build trust and credibility, essential assets for any organization in the modern business landscape [1].

"What gets measured gets managed"

Effective measurement and management practices can cover a broad spectrum of indicators, from carbon emissions and energy efficiency to supply chain logistics, project delivery and resource utilization among many others. Metrics lay a quantifiable foundation for strategic decisions, enabling sustainability initiatives to be rigorously assessed, fine-tuned, and optimized for enhanced outcomes.

Effective measurement systems also catalyze innovation by setting ambitious benchmarks. This enables organizations to challenge their norms and propel technological advancements, leading to novel technologies and processes that elevate the efficiency and sustainability of energy systems.

Applying Drucker's principle in today's energy sector requires broadening the scope of metrics to include sustainability measures. Alongside traditional metrics like cost, operational performance and output, integrating measures, such as renewable energy integration, waste management efficiency, and water usage is crucial. These metrics should also extend to evaluating the social impact of energy production, which encompasses community engagement and employee welfare.

Energy organizations are now increasingly leveraging advanced data analytics and Internet of Things (IoT) technologies to manage these complex metrics. These technologies provide extensive real-time data, enhancing transparency and enabling dynamic management practices that adapt swiftly to new challenges.

While the advantages of implementing careful systems for performance measurement are evident, it presents challenges, such as ensuring data quality, maintaining the integrity of measurement processes, and aligning metrics with strategic goals. Additionally, it is important to balance the cost-effectiveness with the value of measurement systems, and especially consideration of the maintenance of measurement systems, i.e., the cost of ongoing support to prevent them from burdening the organization's resources and overhead costs.

In essence, adopting "what gets measured gets managed" in the energy transition is not solely about compliance; it fundamentally involves building a resilient and sustainable business ready to thrive in a regulated, complex environment. Energy companies excelling in crafting robust measurement and management systems are poised to lead the industry toward a sustainable future, continuously challenging and redefining industry norms.

Case study: Enel Group's journey toward operational efficiency and sustainability

The Enel Group, an Italian multinational energy company and a leading integrated operator in the global power and renewables markets, offers a compelling example of successful implementation of measurement strategies to enhance operational efficiency and sustainability. As the world shifts toward renewable energy and sustainable practices, Enel faced the significant challenge of aligning its vast operations with stringent sustainability goals. The company needed to improve its energy efficiency, reduce carbon emissions, and ensure these metrics could be effectively measured and managed across its global operations.

Enel's approach to tackling this challenge was comprehensive, involving the deployment of advanced digital technologies and the establishment of clear, measurable sustainability targets. Central to this strategy was the incorporation of digital technology into its operations, particularly through the widespread implementation of smart meters. As of 2020, Enel managed over 44 million smart meters in Italy, Spain, and Latin America. These smart meters provided detailed data on energy consumption patterns, crucial for both consumers and the company to understand and optimize electricity use. This digital transformation facilitated predictive maintenance and better network management, leading to fewer outages and disruptions.

A key component of Enel's strategy was its commitment to increasing renewable energy capacity. To measure progress toward this goal, Enel developed specific metrics to track the installation and productivity of renewable energy sources, including wind, solar, hydro, and geothermal installations. These metrics also tracked the reduction in carbon emissions achieved through renewable sources. By the end of 2020, Enel's installed renewable capacity had grown substantially, positioning it as one of the world's leading producers of renewable energy.

Another critical aspect of Enel's measurement strategy was its goal to achieve full decarbonization by 2050. To reach this goal, Enel implemented a detailed system to track its carbon footprint across all operations. This system adhered to international standards and provided crucial data for internal decision-making and external reporting. Accurate carbon footprint measurement allowed Enel to identify key areas for improvement, aligning its operations with global sustainability standards and enhancing its environmental responsibility. Enel also focused on optimizing resource allocation through the use of performance indicators to assess the efficiency of its power plants, logistics of resource distribution, and operational costs associated with generating and distributing power. These measurement strategies enabled Enel to allocate resources more effectively, ensuring optimal use of assets and reducing operational inefficiencies.

The results of these measurement strategies have been significant. Enel reported notable improvements in operational efficiency and a reduction in energy waste. The data from smart meters enabled more efficient network management and predictive maintenance, reducing the frequency and impact of outages. Enel's focus on renewable energy has paid off, with substantial growth in installed renewable capacity by the end of 2020. This growth has made Enel one of the world's leading producers of renewable energy, demonstrating its commitment to sustainable energy practices.

Additionally, Enel has successfully reduced its CO_2 emissions per kWh of electricity generated, aligning with its goal of full decarbonization by 2050.

The Enel Group's case exemplifies how effective measurement strategies can significantly enhance a company's ability to manage its operations sustainably. Setting clear metrics and leveraging advanced technology enabled Enel to improve its operational efficiency and also position itself as a leader in the transition to sustainable energy. This case study underscores the essential truth of Peter Drucker's maxim, "what gets measured gets managed," demonstrating its applicability in driving the energy transition forward [2].

Cultivating a culture of measurement

In the ever-evolving landscape of modern industries, especially those as pivotal as the energy sector, the transition toward data-driven continuous improvement is essential. As we discussed in Chapter 5, this approach is not merely a strategic advantage but a fundamental necessity for surviving and thriving in competitive markets. While deploying cutting-edge technology and methodologies forms the backbone of effective measurement strategies, the true determinant of success lies within the organizational culture. A culture values data-driven insights and continuous feedback is crucial, building a data-driven culture requires the following building blocks.

1. Leadership commitment

The effectiveness and ultimately success of data-driven strategies significantly depends on the commitment from the organization's leadership. Leaders must embody and champion the principles of data-driven management. This involves more than just approving tools and technologies, and it requires active involvement in data-driven initiatives. Leaders should regularly interact with the data, integrate insights into strategic decisions, and communicate the importance of these practices throughout the organization. Leaders need to set a powerful example that encourages all employees to embrace and prioritize data-centric methods.

2. Employee engagement

Employee engagement is critical to the success of data-driven methodologies. Comprehensive training programs are essential to equip employees with the necessary skills to interpret and utilize data effectively. Employees should be involved in setting and tracking performance metrics, which helps them see how their actions directly contribute to the organization's goals. This engagement fosters a sense of ownership and responsibility, crucial for a proactive organizational culture. Moreover, engaging employees in this way ensures that data-driven strategies are understood and embraced across the company, not just confined to data specialists or upper management.

3. Feedback mechanisms

Robust feedback mechanisms are the cornerstone of continuous improvement. Regularly scheduled reviews and feedback sessions provide essential platforms for discussing performance data and improvement strategies. These interactions should be transparent and inclusive, allowing employees at all levels to voice their concerns, suggest improvements, and share insights. Such openness improves the strategies themselves and their effective implementation, and also builds a supportive culture that values each employee's input.

Ultimately, effective feedback mechanisms enhance strategic agility, allowing the organization to adapt quickly to new information or market changes. We will present the idea of a "war room" later in this chapter, which provided a strong vehicle for feedback and action planning.

Overcoming challenges in implementation

Implementing effective measurement strategies in the complex environment of an energy company is fraught with challenges. Common obstacles include resistance to change, the complexity of data management, significant initial investments for technology and training, and various organizational, technical, and operational hurdles.

Effective change management is crucial for mitigating resistance and facilitating the adoption of new technologies and processes. Clear communication about the rationale behind the changes, the benefits expected, and the impact on employees' daily tasks is essential. Transparency builds trust and reduces resistance. Comprehensive training programs, including hands-on workshops, e-learning modules, and continuous learning opportunities, ensure all employees are confident in using new systems and understand the benefits these changes bring. Engaging key stakeholders early in the process to gather input and build support, appointing change advocates or change managers within the organization, and continuously monitoring and adjusting the implementation process based on feedback are essential steps.

Introducing new technologies in phases can minimize disruption and provide time for necessary adjustments. Starting with pilot projects to test new technologies on a smaller scale allows the organization to identify potential issues, make adjustments, and build a success story that can be shared across the company. Implementing changes in smaller teams or departments before a full-scale rollout allows for iterative learning and adjustment, which is essential for managing the complexities of new technology integration. Establishing robust feedback mechanisms to gather input from users during each phase of the rollout and choosing scalable solutions that can grow with the organization are also critical.

Managing and integrating vast amounts of data is another major challenge for organizations. Effective data management strategies should include strong data governance policies to ensure data quality, security, and compliance, as well as integrated systems that can seamlessly share and analyze data across different departments and functions. Utilizing advanced analytics and artificial intelligence to derive actionable insights from the data is also essential. Ensuring organizational alignment is crucial for the successful implementation of measurement strategies. This involves aligning measurement strategies with the organization's overall strategic goals, fostering cross-functional collaboration to ensure different departments work together toward common goals, and securing strong support from leadership to drive the implementation process.

Developing a culture that deeply integrates data-driven continuous improvement is an ongoing endeavor requiring commitment, strategic foresight, and adaptability. Engaging employees at all levels in the continuous improvement process, recognizing and rewarding those who actively contribute to the success of the measurement strategies, and promoting a culture of continuous learning and development are key steps.

Are you measuring the right things?

In the previous section, we explored The Paradigm of Measurement and Management, delving into how the evolution of measurement practices has shaped organizational success. Now, as we challenge the norm, we must ask a fundamental question: Are you measuring the right things? This section builds on the concepts of data-driven continuous improvement and the strategic imperative of embracing digital technologies discussed in Chapters 4 and 5.

To illustrate concept there is an interesting story during World War II, where Allied forces were faced with significant losses of bomber aircraft. The military sought to reinforce the aircraft to reduce these losses but faced a dilemma: adding armor indiscriminately would make the planes too heavy to be effective. The challenge was to determine which parts of the aircraft should be reinforced to maximize protection while minimizing additional weight. Abraham Wald was a renowned statistician and member of the Statistical Research Group at Columbia University during World War II. His innovative work during the war provides a profound lesson in innovative thinking and statistical analysis applied to real-world problems and helps us focus on measuring the right things.

Allied forces sought to reinforce their bombers to reduce losses, initially military analysts focused on the bombers that returned from missions, meticulously recording the locations of bullet holes and other damage. They noted that certain areas, such as the wings and tail, were more frequently damaged and intuitively suggested reinforcing these areas.

However, Wald recognized a critical flaw in this analysis: survivorship bias. This bias occurs when only the successes in this case, i.e., the returning bombers are considered, ignoring the failures, i.e., the bombers that were shot down. Wald's insight was that the bombers that returned with damage were able to survive despite being hit in those areas. Therefore, the undamaged areas on the surviving bombers represented the parts where hits would likely be catastrophic, causing the plane to be lost. He proposed reinforcing these less visibly damaged areas, such as the engine and cockpit, which were critical for the aircraft's survival. This counterintuitive yet statistically sound solution proved effective, reducing bomber losses and improving mission success rates.

Wald's approach has far-reaching implications for modern measurement strategies. It illustrates the importance of looking beyond the obvious and questioning initial assumptions. In contemporary organizational contexts, this means developing measurement strategies that consider both visible indicators and underlying factors that may not be immediately apparent. For example, in an energy company, while tracking metrics like production rates and compliance figures is essential, it is equally important to consider less obvious indicators, such as near-misses, minor anomalies, and underlying operational inefficiencies. Advanced data analytics and machine learning can play a crucial role in uncovering these hidden insights, enabling organizations to develop more comprehensive and effective measurement strategies. Wald's work highlights the value of questioning assumptions, leveraging data-driven insights, and focusing on both visible and hidden factors to drive continuous improvement and long-term success (Fig. 8.3).

Survivorship bias occurs when we focus solely on the survivors of a process, ignoring those that did not make it. This is an important concept and can be applied to a wider range of organizations within the energy sector.

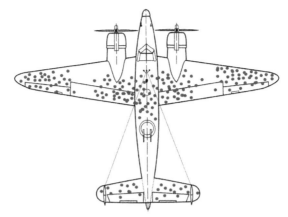

Fig. 8.3 Survivorship bias in WWII bombers.

For example, in project delivery this bias can manifest in several critical ways. Organizations often analyze only the successful projects, those completed safely, on time, within budget, and meeting quality standards. This creates an overly optimistic view, neglecting the challenges and risks inherent in the projects that failed or faced significant issues. By only looking at the "survivors," organizations miss out on valuable lessons that could be learned from the "casualties." Focusing solely on successful projects can lead to several pitfalls. One major issue is overlooking the root causes of failure in other projects. These root causes can range from inadequate risk management, poor stakeholder communication, and insufficient resource allocation to flawed project planning. It is important to analyzing these aspects so that organizations can avoid repeating the same mistakes.

Another pitfall is creating a false sense of security. Analyzing only successful projects can lead organizations to believe that their current processes and strategies are foolproof, ignoring the inherent variability and uncertainty in project execution. This complacency can result in unpreparedness when unexpected challenges arise in future projects. Moreover, successful projects often benefit from favorable conditions that may not be present in every project. Organizations that do not examine projects that faced difficulties or failed might ignore the full spectrum of risks they could encounter, which can lead to inadequate risk management strategies that do not account for worst-case scenarios.

Examining failures is crucial for uncovering hidden lessons. Projects that failed or encountered significant issues often highlight vulnerabilities and weak points in project management practices. For example, a project that failed due to poor supplier management can reveal the importance of stringent supplier evaluation and ongoing performance monitoring of suppliers. Studying failures also helps organizations build resilience. Understanding what went wrong in past projects enables teams to develop contingency plans and mitigation strategies for potential future challenges. This proactive approach helps create a more robust project management framework.

A balanced perspective, which includes analyzing both successes and failures, ensures a comprehensive understanding of project dynamics. This approach helps in creating realistic benchmarks and performance metrics that reflect the true complexity and risks of project delivery. Just as Wald focused on undamaged areas of bombers, project teams should scrutinize aspects of project management, quality control, safety, and procurement that are often overlooked. For instance, examining the quality of supplier materials and the robustness of safety protocols, even if past projects did not face issues in these areas, can preempt potential problems. Addressing the less damaged areas early in the project lifecycle can prevent future failures. If past projects succeeded despite inadequate supplier quality checks, implementing rigorous quality control from the start can mitigate risks. This proactive approach ensures that weaknesses are addressed before they can lead to significant issues. An effective startup process should foster a balanced perspective, combining optimism with a realistic assessment of potential risks. This balanced view encourages teams to learn from both successes and failures, ensuring that lessons from problematic projects are not ignored.

Consider another example for energy companies that operate large-scale assets, such as offshore oil platforms or renewable energy farms. Measuring the right things is obviously crucial for operational excellence. Consider a company managing a fleet of offshore wind turbines. Traditionally, organizations might focus on obvious metrics like power output, turbine uptime, and maintenance costs. While these are important, they may not provide a complete picture of operational health. By focusing solely on these metrics, organizations might overlook less apparent issues that could significantly impact long-term performance. For instance, they might miss subtle patterns in equipment wear and tear, environmental impact on turbine efficiency, or minor operational inefficiencies that, if left unaddressed, could escalate into major problems. A holistic measurement approach is needed. For example, incorporating detailed environmental data into performance metrics can reveal how changes in wind patterns, sea temperatures, and marine growth affect turbine efficiency. This insight allows for proactive adjustments in maintenance schedules and operational strategies.

Leveraging digital technologies and data analytics enables organizations to implement predictive maintenance programs, such as monitoring vibration patterns, temperature fluctuations, and other sensor data so that potential failures can be predicted before they occur, reducing downtime and maintenance costs. Beyond uptime and power output, measuring operational efficiency involves analyzing the entire value chain. This includes logistics for maintenance crews, supply chain efficiency for replacement parts, and even the impact of human factors on performance.

Organizations that adopt this comprehensive measurement approach are more likely to identify and mitigate risks early, optimize operations, and enhance overall asset performance. As highlighted in Chapter 4, embracing digital technologies is not just about having the right tools but also about fostering a culture of continuous improvement and measurement. Organizations must cultivate an environment where data-driven insights are valued, and where teams are encouraged to look beyond the obvious metrics. This cultural shift ensures that the right things are measured, leading to more informed decision-making and better project outcomes.

Survivorship bias can significantly impact project execution. By challenging the norm and asking whether we are measuring the right things, organizations can adopt a more comprehensive and balanced approach, which involves examining overlooked areas, preventing future failures, and integrating data-driven insights to foster continuous improvement. Organizations can learn from past mistakes and ensure a more robust and resilient approach to project management by adopting this approach. This holistic perspective is crucial for achieving operational excellence in the energy transition and digital age.

Building on Chapter 5's discussion on data-driven organizations, it is important to harness digital tools to collect, analyze, and act on data from a wide range of sources. Embracing digital technologies allows for more comprehensive data collection and analysis, ensuring that measurements encompass all critical areas, not just the most visible or successful ones turning data into actionable insights.

Turning data into actionable insights

The energy sector, with its rapid evolution toward sustainability energy and digitization, demands robust strategies that turn extensive data into actionable insights. Leveraging actionable insights derived from data is about optimizing current operations but more fundamentally it is about challenging and transforming standard practices. Actionable insights empower energy companies to make informed, strategic decisions that can significantly alter their operational landscape, enhance sustainability, and drive innovation.

What are actionable insights?

Actionable insights refer to the valuable information extracted from data that companies can use to make decisions that have a direct impact on their performance and strategic direction. These insights provide a deeper understanding of operational efficiencies, market dynamics, customer behavior, and environmental impact. They go beyond mere data collection to interpret and suggest the best courses of action based on real-time and historical data analyses, some examples that help to illustrate this are as follows.

One of the primary areas where actionable insights can transform the energy sector is in enhancing operational efficiency and sustainability. For example, data collected from sensors and IoT devices on energy consumption and output can reveal inefficiencies in real-time. Energy companies can use these data to adjust their operations dynamically, reducing waste and optimizing resource use. Moreover, insights into carbon emissions and other environmental impacts enable companies to implement strategies that not only comply with regulations but also set new standards for environmental stewardship.

Actionable insights can also revolutionize maintenance strategies. Predictive analytics utilize data from equipment sensors to anticipate failures before they occur, allowing for timely maintenance that can prevent costly downtimes and extend equipment life. This proactive approach challenges the traditional reactive maintenance paradigm, leading to more reliable operations and reduced operational costs.

The insights gained from comprehensive data analysis can lead to innovations in how energy is produced and distributed. For instance, understanding patterns in energy demand through data analytics allows for the development of smart grids that more efficiently distribute energy based on real-time demand. This not only improves the reliability of energy supply but also supports the integration of renewable energy sources, which can be more variable in output.

Regulatory compliance is a significant concern for energy companies. Actionable insights can help ensure that companies not only meet existing regulatory standards but also anticipate future regulations. Enhanced data reporting tools can automate the process of gathering and submitting regulatory data, reducing the risk of compliance issues. This proactive approach can set new norms in regulatory adherence, showcasing a company's commitment to transparency and ethical practices.

While the advantages of leveraging actionable insights are clear, there are significant challenges that energy companies must navigate to fully realize these benefits. One major challenge is data overload. Energy companies often collect vast amounts of data from numerous sources, such as sensors, operational logs, market data, and customer interactions. This sheer volume of data can overwhelm traditional analysis methods, making it difficult to extract relevant and actionable insights. In this context, implementing advanced data management and analysis systems becomes crucial. Advanced data management and analysis systems for an energy company typically include several key components. First, robust data storage solutions, such as cloud-based platforms, are essential for handling large volumes of data. These platforms offer scalable storage that can grow with the company's needs and ensure data are securely stored and easily accessible.

Next, data integration tools are necessary to combine data from disparate sources into a unified framework. These tools can harmonize data formats and ensure consistency, making it easier to perform comprehensive analyses. For example, integrating data from field sensors with market price data can provide a holistic view of operational efficiency and financial performance.

Once the data are stored and integrated, advanced analytics tools, such as machine learning algorithms and AI models, come into play. These tools can sift through large datasets to identify meaningful patterns and trends that might be missed by human analysts. Machine learning models, for instance, can predict equipment failures by analyzing historical performance data and identifying early warning signs of potential issues. AI can also optimize energy usage by dynamically adjusting operations based on real-time data inputs.

Visualization tools are another important component of advanced data analysis systems. These tools translate complex data analyses into intuitive visual formats, such as dashboards and interactive charts, that decision-makers can easily understand and act upon. For example, a dashboard displaying real-time energy production and consumption metrics can help managers quickly identify inefficiencies and take corrective action.

Implementing these advanced systems also necessitates a shift to a data-driven culture, which requires substantial change management efforts. Employees at all levels need to be trained in the new technologies and methodologies to effectively use these tools. This involves technical training, as well as adopting a mindset that values data-driven decision-making. Training programs might include hands-on workshops, e-learning modules, and real-time support from data specialists. Appointing data champions within different departments can help bridge the gap between technical experts and operational staff, providing ongoing support and promoting best practices.

Understanding and acting on actionable insights allows organizations to challenge the norm and redefine industry standards. Implementing strategies based on real-time or near-real-time data, organizations can enhance operational efficiencies, drive innovation, improve sustainability, and set new benchmarks in regulatory compliance. However, to truly capitalize on these benefits, energy companies must also address the challenges associated with advanced data practices, ensuring robust data management, and fostering a culture that embraces continual learning and adaptation. As the energy sector continues to evolve, those who effectively leverage actionable insights will lead the transformation, setting new standards for the rest of the industry to follow.

"War rooms": Enhancing operational efficiency

In the context of operations management, the traditional concept of a "war room," also known as a situation room, serves as a dedicated space for strategic decision-making. These rooms are particularly crucial during high-stakes scenarios, such as crisis management, major projects, or critical system overhauls, but they can also be highly effective for day-to-day operations management. War rooms are often equipped with technology to display a vast array of real-time data, from global energy markets to local production statistics, to project delivery dashboards. Large screens may show live data feeds, while interactive dashboards allow users to manipulate data to model different scenarios or outcomes.

Depending on the specific needs of the operation, war rooms may well be equipped with more traditional tools like whiteboards, traffic light colored magnetic pins, and post-it notes to facilitate dynamic planning and problem-solving (Fig. 8.4).

Fig. 8.4 War room example, showcasing agile management.

The evolution of war rooms

Over the years, war rooms have evolved radically with the integration of agile and visual management techniques. When implemented effectively, war rooms can positively impact an organization's success by driving efficient and effective operations management processes or project delivery. They achieve this by connecting data insights with operations teams and empowering decision-makers with real-time information.

Setting up such command centers provides real-time operational oversight and faster decision-making capabilities. These centers utilize up-to-the-minute data visualizations and agile management practices to turn poorly performing areas into efficient operations, significantly enhancing site performance. This proactive approach enables organizations to address issues as they arise and make informed decisions that directly impact productivity and operational efficiency. The setup of a war room encourages collaborative problem-solving and rapid decision-making. Teams from various departments, such as operations, finance, logistics, and compliance, can come together to monitor events as they unfold and respond in real-time. For instance, during an unexpected disruption in the supply chain, the team can quickly gather in the war room to assess the impact using real-time data and coordinate a strategic response to minimize operational downtime and financial loss.

War rooms facilitate long-term strategic planning. Regular meetings held in this setting can focus on forward-looking analytics, such as predictive maintenance scheduling or optimization of energy distribution in response to forecasted demand changes. The centralization of critical information and decision-making processes in the war room enables a unified approach to tackling both immediate and future challenges.

Key tools for war rooms

Stand-up meetings are a cornerstone of effective war rooms, facilitating quick and efficient communication among team members. These daily gatherings are short, focused, and typically held in front of a task board to ensure everyone is aligned and aware of the current status of tasks. Participants stand to encourage brevity and attentiveness. During a stand-up, each team member provides a brief update on their tasks, focusing on what they accomplished yesterday, what they plan to do today, and any obstacles they are facing. This format ensures that everyone is aligned, aware of current priorities, and can quickly address any issues that might impede progress. These meetings need to be carefully managed with strict adherence to agendas and timing, and to ensure the right level of focus is given on each task to maintain the operating rhythm. Typically, these meetings are limited to 15 min to maintain efficiency and focus. By holding these meetings daily, the team can maintain a clear and up-to-date picture of project progress and quickly adapt to any changes or challenges that arise.

In a well-functioning war room, it is crucial to separate status updates from in-depth problem-solving discussions. Stand-up meetings should be dedicated to quick updates, while any identified problems or challenges that require deeper discussion are scheduled for separate problem-solving sessions. This approach ensures that stand-ups remain focused and efficient, while allowing time for detailed analysis and resolution of issues in dedicated sessions. Focusing on actionable items during stand-ups and reserving detailed problem-solving for separate meetings prevents the stand-up from becoming bogged down in lengthy discussions.

Maintaining a high pace in meetings is important to keep the team focused and on track. To achieve this, often a timekeeper can be appointed who is responsible for monitoring the meeting duration and ensuring that discussions do not exceed the allocated time. Predefining the agenda and assigning specific time slots for each topic can help manage the pace and prevent overruns. This approach keeps meetings dynamic and ensures that all necessary points are covered within the scheduled time.

Punctuality is critical in maintaining the discipline and effectiveness of war room meetings. Enforcing strict attendance rules emphasizes the importance of being on time and shows respect for everyone's schedules. Latecomers should be excluded from the meeting to reinforce the importance of discipline. This practice although seems harsh is very effective and helps maintain a high level of engagement and commitment from all team members, ensuring that meetings start and end on time and that discussions are not interrupted by late arrivals.

Effective war rooms bring together diverse teams from various functional areas, such as operations, finance, logistics, and compliance. This diversity of expertise and perspectives is crucial for comprehensive problem-solving. By involving members from different functions, the team can leverage their collective knowledge to develop well-rounded solutions and address issues from multiple angles. This collaborative approach enhances the quality of decision-making and ensures that all relevant factors are considered.

Task boards are a vital tool in war rooms, providing a visual management system that helps teams track progress and stay organized. These boards are customized to align with specific processes and project objectives. They visually display tasks, priorities, and progress, helping the team identify bottlenecks and focus on critical areas. Regular updates to the task board ensure that it reflects the current status of tasks, providing a clear and up-to-date picture of project progress.

Encouraging active participation is key to the success of war room meetings. One way to achieve this is by rotating the facilitator role among team members. This practice encourages ownership and brings diverse leadership perspectives to the meetings. Emphasizing shared priorities and collective goals ensures that all team members are actively engaged in discussions and decision-making. Active participation fosters a sense of responsibility and commitment to the team's objectives.

Continuous improvement is an essential principle in effective war rooms. Regularly identifying and documenting best practices and lessons learned helps the team continuously refine their processes. Assigning specific improvement projects to team members ensures that these practices are implemented and that the team remains focused on enhancing efficiency and effectiveness. Creating a feedback loop allows team members to provide input on war room activities and suggests areas for improvement, fostering a culture of continuous enhancement.

Visual management

Visual management leverages tools, such as dashboards, real-time performance metrics, and infographics, strategically placed around work areas or displayed on digital platforms accessible to all team members. This method helps demystify complex data and metrics, making them accessible and understandable to employees at all levels of the organization.

In addition to standard dashboards, which are central to visual management by providing a real-time overview of KPIs and consolidating data from various sources into a user-friendly format, several other techniques can enhance visual management in war rooms, some useful approaches are as follows:

Kanban boards

These boards help visualize workflow and manage tasks efficiently. Teams can easily track progress and identify bottlenecks by displaying tasks in columns that represent different stages of the process (e.g., "To Do," "In Progress," "Done"). Kanban boards promote transparency and provide a clear view of task statuses, helping teams prioritize work and allocate resources effectively.

Gantt charts

Gantt charts are useful for project delivery and show a timeline of tasks and their dependencies. These charts help teams plan and coordinate activities, supporting projects to stay on schedule. Visualizing the start and end dates of tasks enables teams to better manage project timelines and anticipate potential delays, facilitating proactive adjustments.

Heat map

These visual tools highlight areas of high activity or concern, such as equipment usage or incident rates. Heat maps help teams focus their attention on critical areas that need immediate action. Displaying data in a gradient of colors allows heat maps to make it easy to identify hotspots and allocate resources where they are most needed.

Infographics

Simplifying complex data into visual summaries, infographics are effective for communicating important information quickly. They can be used to display safety protocols, environmental impacts, or performance summaries. Infographics make data more digestible and engaging, helping team members quickly grasp key insights and take informed actions.

Benefits of visual management

Visual management enhances transparency and fosters a culture of accountability and continuous improvement by making data visible and accessible. Employees are then able to see the impact of their work in real-time and understand how their efforts contribute to the organization's goals. This visibility encourages proactive problem-solving and supports a collaborative environment where everyone is aligned with the strategic objectives.

Beyond operational data, visual management can also reinforce corporate values and objectives related to sustainability and community engagement. For instance, displays of monthly or yearly progress toward reducing carbon emissions, water usage, or contributions to local communities can motivate employees to participate actively in these initiatives. Making such information visible enables companies to reinforce the importance of these goals and encourage a collective effort to achieve them. Visual management tools can improve operational efficiency and help build a cohesive team environment. When employees see the direct impact of their actions on overall performance, it fosters a sense of ownership and responsibility. Moreover, visual tools can facilitate better communication and collaboration by providing a common reference point for discussions and decision-making. This, in turn can lead to more effective problem-solving and innovation.

Integrating various visual management techniques, such as Kanban boards, Gantt charts, heat maps, and infographics, is a tried and tested way to enhance operational efficiency and effectiveness. These tools provide clarity, improve communication, and help teams stay focused on their goals, ultimately driving better performance and continuous improvement. Couple with the technology that the digital era has to offer, such as integrated dashboards and actionable insights that are generated from AI, can even further enhance operational efficiency, and effectiveness.

Agile management is a dynamic and flexible approach to project management that emphasizes iterative progress, collaboration, and adaptability. This methodology aligns seamlessly with the key tools and practices discussed previously, such as stand-up meetings and visual management. Stand-up meetings, a staple in agile practices, foster quick and efficient communication, ensuring teams are aligned and obstacles are promptly addressed. Similarly, visual management tools like Kanban boards and Gantt charts are central to agile management, providing real-time visibility into project progress and facilitating proactive adjustments. Integrating agile principles into war rooms can further enhance their efficiency, responsiveness, and ability to deliver high-quality results.

Agile management

Agile project management is a relatively new concept that has become prominent with the rise of software development. When comparing agile project management to traditional project management, the comparison is primarily between agile and the "waterfall" methodology.

The waterfall methodology is a framework for delivering projects in a linear, sequential manner. Unlike agile, which operates in fast, iterative cycles, waterfall projects progress from one phase to the next only after completing all tasks in the current phase. For example, all requirements for the entire project are gathered before moving on to the design phase. The waterfall methodology, depicted by frameworks, such as PMP and PRINCE2, was developed and implemented in the late 1900s, primarily for construction and manufacturing projects. These projects required tight control, and a linear waterfall approach helped minimize risk and clearly define the project scope. In the energy transition era, often with uncertain or evolving work scopes, a more flexible approach may well be required and agile offers a compelling option (Fig. 8.5).

Agile management techniques have transformed project management and operations by prioritizing flexibility, responsiveness, and iterative progress, they are particularly suitable for innovative projects. These techniques emphasize adaptability, collaborative effort, and a strong focus on delivering customer value. Here, we explore the key principles of agile management, enriched with relevant theories and practices, and their relevance to challenging traditional norms in operational excellence.

A core principle of agile management is iterative development, which breaks projects into smaller, manageable parts known as sprints. Short, time-boxed periods during which specific work is completed and made ready for review. Sprints enable teams to focus on small, achievable goals and adjust plans based on feedback. Sprints allow for regular reassessment and adaptation based on feedback and changing conditions. Some of the

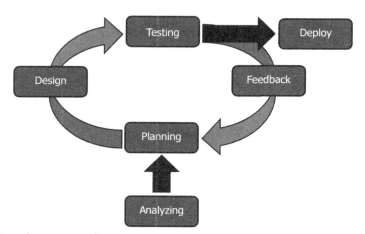

Fig. 8.5 Agile project cycle.

key concepts are summarized next. A Sprint Review can be held at the end of each sprint to demonstrate accomplishments. Stakeholders provide feedback, guiding future work. A Sprint Retrospective is a reflective meeting where the team discusses successes, challenges, and process improvements for the next sprint.

Collaboration is a cornerstone of agile management, emphasizing teamwork and open communication across all organizational levels. Regular meetings and updates ensure that everyone is informed and aligned, breaking down silos and fostering a culture of collective problem-solving. Techniques, such as daily stand-ups, cross-functional teams, and virtual collaboration tools, are integral to this principle as discussed previously.

A strong customer focus is central to agile management, with teams prioritizing customer needs and feedback to ensure the end product or service meets requirements. This customer-centric approach drives continuous improvement and innovation. Key components include user stories, customer feedback loops, and the product owner representing the customer's interests.

Agile management emphasizes empowerment, with teams given the autonomy to make decisions and solve problems, fostering ownership and accountability. Self-organizing teams, decentralized decision-making, and continuous learning and improvement are crucial elements.

Implementing agile management techniques offers numerous benefits, including increased flexibility, enhanced collaboration, greater customer satisfaction, and higher productivity. Agile methods allow teams to adapt quickly to changes in market or project requirements, while regular communication and collaboration improve teamwork and problem-solving. Focusing on customer needs ensures that agile teams deliver products that better meet expectations, and empowered teams are more motivated and efficient, driving innovation.

Agile management techniques, grounded in iterative development, collaboration, customer focus, and empowerment, provide a robust framework for managing complex projects and operations. Integrating agile practices into war rooms and other operational settings amplifies these benefits, fostering a dynamic and high-performing environment. This approach aligns with the concept of challenging the norm, encouraging organizations to continuously refine their processes, embrace change, and deliver greater value to their customers. Adopting agile management principles enable organizations to enhance their responsiveness and drive continuous improvement, especially to the challenges of the ET.

Case study: Implementing "basecamp" in a project services company

In the contemporary landscape of project services, managing multiple projects effectively while ensuring high standards of performance can be a major challenge. Kent plc, a world class project services company, embarked on a transformative journey to enhance its operational excellence through the implementation of a dedicated war room, aptly named "Basecamp."

This initiative is designed to leverage the principles of visual management, agile methodology, and continuous improvement to optimize project delivery and at the same time promote a culture of transparency and accountability.

The primary goal of Kent's Basecamp is to provide a structured environment where key metrics and performance indicators can be monitored and discussed regularly, and prompt action taken. This initiative aimed to streamline operations, enhance communication across departments, and ensure that all team members are aligned with the company's strategic objectives. By implementing Basecamp, Kent sought to:

1. Improve operational control and efficiency by providing real-time visibility into project performance and resource allocation.
2. Enhance communication by facilitating regular, structured meetings that ensure all stakeholders are informed and engaged.
3. Drive continuous improvement by using data-driven insights to identify and address operational bottlenecks and inefficiencies.

The physical layout of the Basecamp room is designed to maximize the use of visual management tools. The room is equipped with multiple visual management boards (VMBs) that display KPIs and project metrics. These boards are updated regularly and serve as focal points during the weekly operational meetings. Innovative solutions, such as smart boards, were utilized to display data effectively. Each VMB includes sections for capturing key metrics, progress updates, and action items. The data are visualized using a traffic light system (green for positive performance, red for areas needing attention) to facilitate quick assessment and decision-making.

Governance is a critical aspect of Basecamp's success. Roles and responsibilities are clearly defined to ensure accountability and effective management of the visual management program, key roles as follows:

Room Owner: Typically, the project or site manager was responsible for maintaining the standards of the Basecamp room and leading the meetings.

VMB Owners: Department heads or designated leaders who are responsible for updating and presenting their respective VMBs.

The meetings are conducted with a strict agenda to maintain focus and efficiency. Key rules include punctuality, respect, and preparedness. The meetings are designed to be dynamic and engaging, with each VMB Owner given a limited time to present their metrics and discuss any red flag items.

The rollout of Basecamp is conducted in phases to ensure a smooth transition and effective adoption across the organization including initial rollout which focuses on key departments and top-level management to set the tone and establish the framework for Basecamp. Full implementation expands the scope to include all relevant departments and integrates the visual management tools into daily operations. Finally, continuous improvement is the ongoing phase where feedback is collected, and the system is refined to enhance performance continually.

The Basecamp program at Kent included the following key elements:

Digital training program: A comprehensive digital training program was implemented for all employees, focusing on data analytics, and digital platforms. This training was instrumental in enabling staff to interpret and utilize the data displayed on the VMBs effectively.

Integration of new roles: New roles, such as Basecamp Manager, VMB Owner, and Continuous Improvement Engineers, were introduced to support the rollout, digital transformation and ongoing support. These professionals played a key role in automating data collection and visualization processes, ensuring that the information on the VMBs was accurate and up-to-date.

Weekly operations meetings: Weekly operations meetings became the cornerstone of the Basecamp initiative. These meetings, attended by department heads and key project managers, focused on reviewing the KPIs, discussing challenges, and identifying improvement opportunities. The structured format and visual aids facilitated productive discussions and quick decision-making.

Lean methodology and continuous improvement: Lean principles were embedded into the Basecamp philosophy, promoting a culture of continuous improvement. Teams were encouraged to identify waste, streamline processes, and implement innovative solutions to enhance efficiency.

Outcomes and benefits

The implementation of Basecamp at Kent resulted in significant improvements across various metrics, including:

- Enhanced operational performance and efficiency, where the structured meetings and real-time visibility into performance metrics enabled quicker response times and more effective resource management.

- Improved communication and collaboration, with regular interactions and transparent sharing of data a collaborative environment where departments worked together toward common goals was developed.
- Continuous improvement, which focused on lean principles and continuous improvement and led to a culture where employees were proactive in identifying and addressing inefficiencies.

The "Basecamp" initiative at Kent is a prime example of how a project services company can leverage visual management and lean principles to drive operational excellence. Creating a dedicated space for monitoring and discussing operational performance enabled Kent to foster a culture of accountability, continuous improvement, and strategic alignment, positioning itself for sustained success in the energy sector [3].

Reflections

In this chapter, parallels have been drawn between the high-stakes world of Formula 1 racing and the energy sector's ongoing transformation, illustrating the necessity of continuous improvement and innovation. The evolution of Formula 1 pit stops from leisurely affairs to under-two-second highly organized events highlights the critical need for energy companies to reengineer their operations to meet the demands of the current energy landscape. Just as Formula 1 teams have leveraged advanced technologies and streamlined processes to gain competitive advantages, energy companies must embrace similar strategies to navigate the challenges posed by rapid technological advances and stringent environmental regulations.

The energy sector's push toward renewable sources, integration of new technologies, and the imperative to reduce carbon footprints demands a paradigm shift in operations. Traditional methods that prioritize stability over speed and innovation are no longer viable. By adopting a mindset of continuous improvement and challenging established norms, energy companies can optimize every facet of their operations from energy production processes to project management and data utilization, thereby enhancing efficiency and sustainability. This proactive approach is essential for companies to survive and thrive in the new era of energy transition.

The concept of continuous improvement, central to this chapter, compels companies to iteratively test, refine, and perfect their methodologies. This framework is crucial for building the agility needed to respond to emerging technologies and shifting market dynamics effectively. The role of CoEs in driving sustained improvement and innovation is pivotal. By centralizing expertise, standardizing processes, and fostering collaborative networks, CoEs enable companies to lead in shaping the future of the industry. This approach institutionalizes the pursuit of excellence and innovation, ensuring that organizations remain at the cutting edge.

Peter Drucker's principle, "what gets measured gets managed," is foundational for driving operational efficiency and transformative change in the energy sector. Effective measurement and management practices are essential for fostering a culture of continuous improvement and strategic decision-making. The case study of the Enel Group exemplifies how implementing sophisticated tracking systems and leveraging advanced technologies can significantly enhance operational efficiency and sustainability. Enel's comprehensive approach to measurement, involving the deployment of smart meters, tracking renewable energy capacity, and optimizing resource allocation, underscores the importance of clear metrics and data-driven insights.

In fostering a culture of continuous improvement, energy companies must overcome challenges such as resistance to change, data management complexities, and the need for significant initial investments. Effective change management, comprehensive training programs, and robust feedback mechanisms are crucial for mitigating these challenges. Embracing a data-driven culture enables companies to ensure that all employees are engaged in the continuous improvement process and that data-driven strategies are understood and embraced across the organization.

As we conclude this chapter, it is clear that challenging the norm and implementing effective measurement strategies are critical for achieving operational excellence in the energy transition era. However, achieving this transformation requires a dedicated and skilled workforce. In the next chapter, we will explore the workforce needed to challenge the norm highlighting the skills, training, and cultural shifts necessary to drive continuous improvement and innovation in the energy sector.

References

[1] P.F. Drucker, The Effective Executive: The Definitive Guide to Getting the Right Things Done, Harper Business, 2006.
[2] Enel Company, Transition to Renewable Energy in Europe: Enel's Strategy and Implementation, 2023.
[3] Kent plc Website, Case Study: Implementing 'Basecamp' in a Project Services Company, 2023.

CHAPTER NINE

Workforce transformation

The energy sector stands on the precipice of a monumental transformation, one that rivals the revolutionary discovery of oil. This transformative shift is driven by the urgent and interconnected needs to combat climate change, enhance energy security, and promote sustainability. The global push toward renewable energy sources, such as solar, wind, and geothermal, is fundamentally reshaping the landscape of energy production, distribution, and consumption. These renewable sources are not just supplementing but are increasingly replacing traditional fossil fuels, heralding a new era of cleaner, more sustainable energy practices.

As we navigate this transition, we witness profound changes across the entire energy value chain. The energy mix is evolving rapidly, with a growing recognition of the need to reduce greenhouse gas emissions and increase energy efficiency. This is further compounded by the imperative to embrace innovative technologies that can drive sustainable energy practices. The digital revolution plays a pivotal role in this transformation, introducing advanced analytics, machine learning, and artificial intelligence (AI) into the industry. These technologies are revolutionizing operations, enabling energy companies to optimize processes, minimize costs, and significantly enhance their environmental performance.

The climate crisis imposes substantial pressure on the oil and gas industry, necessitating a reduction in greenhouse gas emissions and a transition toward more sustainable energy sources. In response, many companies are diversifying their portfolios, investing heavily in renewable energy projects, and exploring cutting-edge technologies, such as carbon capture and storage. This diversification is crucial not only for environmental sustainability but also for maintaining economic viability in an increasingly carbon-constrained world.

However, it is important to recognize that this transition will not happen overnight. Fossil fuels will continue to play a role in the foreseeable future, but the trajectory is clear the energy sector is undergoing an irreversible transformation. New opportunities are emerging for companies that are ready to adapt, innovate, and lead. Conversely, those that fail to embrace these changes, or adapt too slowly, will face significant challenges and potential obsolescence.

Powering Through the Transition
https://doi.org/10.1016/B978-0-323-91754-4.00009-1

In this rapidly evolving landscape, agility and proactivity are essential. Companies must be willing to reinvent themselves, acquiring new skills and forging partnerships within entirely new ecosystems. Successfully navigating this transformation requires a robust embrace of innovation and a keen pursuit of new opportunities. This may involve exploring novel business models, investing in groundbreaking technologies, and adopting fresh approaches to collaboration and partnerships.

Ultimately, the goal is to create a more sustainable, efficient, and technologically advanced energy industry capable of meeting the demands of future generations. This chapter delves into the critical role of the workforce in this transition, examining the skills, roles, and cultural shifts needed to thrive in the new energy era. It explores how companies can leverage digital advancements and data-driven strategies to empower their employees and drive sustainable growth.

The digital revolution and workforce transformation

The digital revolution has profoundly reshaped the energy sector, ushering in an era characterized by unprecedented efficiency and sustainability. At the heart of this transformation are technologies like AI, machine learning, advanced analytics, and the Internet of Things (IoT). These innovations are no longer mere concepts of the future, they are vital tools driving today's industry operations, enabling companies to optimize processes, reduce costs, and significantly improve environmental performance. However, the successful implementation of these technologies depends on a workforce equipped with the necessary skills and adaptability to keep pace with the continuous evolution of digital tools.

Traditional roles are being redefined, and new roles are emerging to facilitate the industry's shift toward the new digital era. Employees now need to be proficient in digital platforms, data interpretation, and continuous learning. For instance, roles that once focused primarily on manual operations now require proficiency in digital platforms and data interpretation. Operators and engineers who traditionally relied on hands-on experience must now integrate digital tools into their workflows. This evolution is necessary to harness the full potential of digital transformation.

The shift toward a data-driven approach necessitates the creation of new roles within energy companies. Data scientists, AI specialists, and digital engineers have become the "new normal." These professionals are tasked with analyzing vast datasets, developing predictive models, and implementing automation technologies that drive efficiency and innovation. Data scientists, for instance, play a crucial role in extracting actionable insights from large volumes of data, enabling companies to make informed decisions.

AI specialists focus on developing and refining machine learning algorithms that can predict equipment failures, optimize production processes, and enhance overall operational efficiency. Digital engineers are responsible for integrating digital technologies into existing infrastructure, ensuring seamless operations and improving data flow across the organization.

To keep pace with rapid technological advancements, fostering a culture of continuous learning is crucial. Energy companies must invest in regular training programs to upskill their workforce. This involves not only technical training in AI, machine learning, and data analytics but also fostering an adaptable mindset that embraces change and innovation. Training programs can include workshops, online courses, and hands-on training sessions designed to enhance employees' technical skills and knowledge.

Investing in workforce development

Energy companies must recognize the importance of investing in workforce development providing employees with the necessary tools and resources to enhance their skills to ensure that their workforce remains capable of meeting the challenges of the digital era. This investment improves operational efficiency as well as enhances employee satisfaction and retention. For instance, companies can establish partnerships with educational institutions to offer specialized training programs tailored to the needs of the energy sector. Additionally, creating mentorship programs and encouraging knowledge sharing among employees can foster a collaborative learning environment.

Embracing change and innovation is vital for workforce transformation. Leaders must encourage a mindset that welcomes new ideas and approaches. This involves creating an organizational culture that values experimentation, supports risk-taking, and rewards innovation. By doing so, companies can cultivate a workforce that is not only skilled but also motivated to learn new skill sets, drive continuous improvement and innovation.

Case study: Shell's digital transformation

Shell's digital transformation journey exemplifies how a traditional energy company can successfully navigate the digital revolution, aligning with the broader context of workforce transformation. Shell has invested in digital skills training and empowered its workforce to leverage AI and machine learning for various applications, demonstrating the critical need for new competencies and roles in the energy sector.

In February 2021, Shell, C3 AI, Baker Hughes, and Microsoft announced the launch of the Open AI Energy Initiative (OAI), a groundbreaking open ecosystem of AI-based solutions designed to transform the energy and process industries. The OAI provides a collaborative framework for energy operators, service providers, equipment providers, and independent software vendors to offer interoperable AI and physics-based models, monitoring, diagnostics, prescriptive actions, and services. This initiative aims to accelerate the digital transformation of the energy sector by combining the expertise of global leaders to promote safer, more efficient, and sustainable energy production. C3 AI CEO Thomas M. Siebel highlighted the initiative's objective to ensure climate security and advance the digital transformation of the industry.

The first OAI solutions, developed by Shell and Baker Hughes, focus on improving the reliability and performance of energy assets and processes. These solutions provide AI-enabled insights to predict performance risks on assets. The Baker Hughes C3 (BHC3) application integrates data from various sources to train AI models covering full plant operations, leveraging Microsoft's Azure cloud infrastructure. This integration highlights the need for new roles such as digital engineers, and data scientists to be incorporated into the workforce to manage and optimize such complex digital systems.

Shell and Baker Hughes are offering several predictive maintenance modules through the OAI, including Shell Predictive Maintenance for Control Valves, Rotating Equipment, and Subsea Electrical Submersible Pumps. These modules illustrate the practical applications of AI and machine learning in enhancing operational efficiency and reliability, reflecting the need for continuous learning and adaptation among the workforce.

The OAI augments BHC3 Applications with partner-led, domain-specific solutions to unlock significant economic value while making energy production cleaner and safer. The reliability solutions include pretrained AI models, failure prevention recommendations, and prescriptive actions, which improve asset uptime and operational efficiency. This approach requires the collaborative and innovative culture that energy companies must adopt to embrace the digital era.

Shell's Chief Technology Officer, Yuri Sebregts, emphasized the practical impact of AI, noting that Shell monitors over 5200 pieces of equipment using machine learning across various assets. This capability has significantly reduced costs and improved the reliability and performance of Shell's operations. Such achievements demonstrate the importance of investing in workforce development to equip employees with the necessary digital skills and knowledge.

The initiative reflects a collaborative approach to leveraging digital technologies for industry-wide benefits. Uwem Ukpong of Baker Hughes and Darryl Willis of Microsoft emphasized the importance of open data standards and AI capabilities in addressing industry challenges like nonproductive downtime. The OAI aims to further the transition to a net–zero emissions future by enhancing plant reliability and maintenance through digital technology. This initiative represents a promising step toward the digital transformation of energy, showcasing the potential of AI to drive significant improvements in operational efficiency and sustainability. This case study stresses the transformative potential of digital technologies in the energy sector and the crucial role of an adaptable, skilled workforce in realizing these benefits [1].

Emerging job roles

The digital revolution is creating entirely new job roles tailored to the evolving needs of the energy sector. These roles are pivotal in enabling companies to harness the power of digital tools and data-driven insights effectively. The traditional energy landscape is being reshaped by advanced technologies such as AI, machine learning, and automation, requiring a workforce adept at navigating these innovations. As companies integrate these technologies into their operations, new roles emerge that blend technical expertise with strategic insight, ensuring that digital transformations lead to tangible operational improvements and competitive advantages.

As the energy sector navigates through this transformative period, it is important to understand and implement these emerging roles to stay competitive and sustainable. The shift toward a more digitized and automated industry is not merely about adopting new technologies, it is about embedding these technologies into the core of organizational strategies and operations. This necessitates a workforce that is not only skilled in the latest digital tools but also adaptable and forward-thinking. Organizations must proactively invest in these new roles to foster a culture of continuous innovation and improvement.

In the context of the energy transition, the need for these new roles becomes even more critical. The transition to renewable energy sources and sustainable practices demands a paradigm shift in how companies operate. Advanced digital technologies can drive this change, but only if there are skilled professionals who can leverage these tools effectively. This transformation requires a holistic approach where new job roles are integrated into the strategic framework of the organization, ensuring that every aspect of the business benefits from digital advancements.

Moreover, the rapidly changing landscape of the energy sector means that companies must stay ahead of the curve by evolving their workforce capabilities. The traditional methods of operation are being replaced by more efficient, data-driven approaches. This shift is not just about technology but also about creating an agile workforce that can adapt to new challenges and opportunities. The emerging job roles in the digital era are designed to meet these needs, driving forward the agenda of sustainability and efficiency.

The importance of these roles cannot be overstated. They are not merely supplementary to the existing workforce but are integral to the core functioning of modern energy companies. As the industry moves toward a more interconnected and digital future, the professionals who fill these roles will be at the forefront of innovation, ensuring that companies can navigate the complexities of the energy transition effectively. Investing in these new roles is a strategic necessity and fundamental requirement for achieving long-term success and sustainability in the energy sector.

Embracing data-driven organizations

As the energy transition and digital era commingle, data have emerged as a critical asset, driving informed decision-making, enhancing operational efficiency, and reducing downtime, a common theme through this book. The shift toward data-driven operations is transforming the way energy companies' function, enabling them to harness the power of big data, advanced analytics, and machine learning to optimize their processes and achieve sustainability goals. However, realizing the full potential of these technologies requires a workforce proficient in data literacy.

To build a data-literate workforce, companies must invest in comprehensive training programs that elevate their employees' skills in data analysis and interpretation. These programs can take various forms, including workshops, online courses, and hands-on training sessions that focus on the practical application of data in everyday operations.

For instance, many of the Big Oil members have implemented a data literacy programs that include a blend of classroom training, e-learning modules, and practical exercises. One of the oil giants, Chevron, has implemented a data literacy program, which covers topics, such as data collection, data cleaning, statistical analysis, and data visualization, ensuring that employees at all levels can leverage data to make informed decisions. Chevron's program is designed to be inclusive, catering to different levels of data proficiency from beginners to advanced users.

The classroom training sessions provide a foundational understanding of data principles and hands-on experience with tools like Excel, Python, and R for data analysis. E-learning modules offer flexible, on-demand learning opportunities that employees can access at their own pace, covering advanced topics like machine learning and predictive analytics. Practical exercises and real-world case studies allow employees to apply their learning to actual business scenarios, fostering a deeper understanding of how data can drive operational improvements and strategic decision-making.

Chevron's commitment to data literacy is part of a broader digital transformation strategy aimed at enhancing operational efficiency and innovation. By equipping its workforce with strong data skills, Chevron can improve individual performance as well as inspiring a culture of data-driven decision-making throughout the organization. This approach helps the company arm its employees with the necessary knowledge and skills needed to stay competitive in the rapidly evolving energy sector and supports it's objective to harness the power of data to achieve its strategic goals [1].

Adopting a data-driven approach requires more than just investing in technology and training, it necessitates a cultural shift within the organization. Companies must foster a culture that values data and encourages employees to leverage data in their daily decision-making processes. Hilary Mason's book, "Data-driven: Creating a data culture" emphasizes the importance of fostering a data-centric culture within organizations. Mason highlights the critical role of data in driving modern business decisions, particularly relevant to the energy sector, which is undergoing significant transformation due to digitalization and the shift toward sustainability.

Mason affirms that a data-driven culture is not merely about having the right tools and technologies but about fundamentally changing the way organizations think and operate. This involves embedding data literacy at all levels, promoting a mindset that values data, and ensuring that data are accessible and usable for all employees.

One of the key takeaways from Mason's work is the necessity of leadership commitment to fostering a data-driven culture. Leaders must advocate for the use of data in decision-making, model data-driven behaviors, and invest in the necessary infrastructure and training, a topic we will dive into in the next chapter, with a strong emphasis is on agile, flexible leadership that embraces innovation and continuous improvement.

Mason also highlights the importance of upskilling employees to handle data effectively. This involves comprehensive training programs that teach not only technical skills like data analysis and visualization but also how to interpret and apply data insights to business decisions, where the focus is on developing competencies in digital platforms, data interpretation, and continuous learning. For example, Chevron's data literacy program, which includes classroom training, e-learning modules, and practical exercises, is an embodiment of the principles Mason advocates. Furthermore, Mason discusses the need for new roles tailored to the data era. These roles are critical in analyzing vast datasets, developing predictive models, and implementing automation technologies that drive efficiency and innovation [2].

The transition to a data-driven organization is a critical component of the energy sector's digital transformation. Investing in data literacy, hiring specialized roles, and fostering a data-driven culture enables organizations to harness the power of data to optimize operations, improve asset utilization, and achieve sustainability goals. Embracing a data-driven approach allows organizations to navigate the complexities of the energy transition more effectively, making informed decisions that drive growth and innovation in a rapidly evolving landscape. This transformation is not just about adopting new technologies but about fundamentally rethinking how the workforce operates and leverages data to drive business success.

Organizations should encourage employees to experiment with data-driven solutions and innovate. This can be achieved in several ways from implementing incentive programs that reward employees for developing data-driven solutions that enhance operational efficiency and sustainability or by creating cross-functional teams that bring together diverse skill sets and perspectives to solve complex problems using data. For example, setting up networks of likeminded employees and providing access to digital tools, encouraging training and development in these areas, the Digital Citizen.

We touched on the concept of the Digital Citizen in Chapter 6, which is a fundamental part of the new-aged workforce required in the energy transition and era. A Digital Citizen is an empowered employee who thrives on innovation, harnesses digital tools, and relies on data-driven approaches to propel the industry forward. There is a clear benefit for energy companies to invest in their Digital Citizens, to drive innovation and efficiency in their organizations. Innovation is at the heart of the Digital Citizen. Encouraging employees to think creatively and explore new ideas, energy companies can unlock unprecedented levels of efficiency and productivity. Digital Citizens are not afraid to challenge the status quo, developing digital solutions to support their organizations, that address the unique challenges of the energy transition.

Equipping Digital Citizens with the right tools is essential. Providing access to advanced digital tools, such as low-code platforms, including Microsoft Power Apps, Power BI or equivalent, and AI tools, encourages employees to get more involved in data management, automating of routine tasks, and creating custom applications to enhance workflows. By democratizing these technologies, organizations ensure that all employees, regardless of their role, can contribute to digital transformation efforts. AI and machine learning can optimize operations and enhance decision-making processes, Power Apps can enable employees to build custom applications without needing extensive coding knowledge, and Power BI can provide powerful data visualization and business intelligence tools to turn data into actionable insights.

Creating networks and communities of practice is vital for nurturing Digital Citizens. These networks facilitate knowledge sharing, best practices, and collaboration, allowing employees to connect with peers, seek advice, and cocreate solutions. Such environments foster continuous learning and innovation, driving the organization forward. Internal communities of practice can form around specific digital skills or technologies, cross-functional teams can leverage diverse perspectives and expertise, and external partnerships with experts, academic institutions, and technology providers can help the organization stay abreast of the latest developments and innovations.

Although Digital Citizens tend to be self-taught through their passion for digital knowledge, continuous training and development are important for the sustainability of the Digital Citizen concept. As the digital landscape evolves, so must the skills of the workforce. Energy companies need to invest in training programs that cover a wide range of digital competencies, including data analytics, machine learning, cybersecurity, and specific digital tools. Regular training opportunities ensure employees stay ahead of the curve and are equipped to tackle the challenges of the digital era. Digital literacy programs can offer foundational training to all employees, advanced technical training can provide in-depth knowledge on topics like AI and data science, and soft skills development can equip employees with essential skills such as critical thinking, problem-solving, and adaptability.

New skills for the energy transition

To be successful in the new energy transition (ET) and digital landscape, energy companies must cultivate a workforce proficient in both digital tools and sustainable practices. This section delves into the key concepts and skills essential, highlighting the roles that will make this transformation possible.

AI and automation

AI and automation are at the forefront of the digital revolution in the energy sector. These technologies enable companies to streamline operations, enhance efficiency, and reduce operational costs. AI algorithms can analyze vast amounts of data to identify patterns and optimize processes, while automation technologies can handle repetitive tasks, freeing up human resources for more strategic activities.

To effectively leverage AI and automation, employees must possess a deep understanding of these technologies. This includes knowledge of AI algorithms, machine learning techniques, and automation tools. Additionally, proficiency in software development, systems integration, and cybersecurity is essential to ensure the robustness and security of digital systems.

Emerging roles, such as Digital Engineers and Automation Engineers, are pivotal in implementing AI and automation technologies. These professionals are tasked with designing, developing, and maintaining digital solutions that integrate seamlessly with existing systems. Their expertise ensures that digital systems are robust, secure, and capable of supporting the company's operations.

Advancing sustainability practices

Sustainability is a central aspect of business strategy in the energy sector. Companies must integrate sustainable practices into their core operations to meet regulatory requirements, satisfy stakeholder expectations, and contribute to global sustainability goals.

Employees need to be well-versed in sustainability principles and practices. This includes knowledge of environmental science, regulatory frameworks, and corporate sustainability strategies.

Sustainability Officers are now essential postings at Energy Company Top Leadership, having critical agendas within the ET. These professionals are responsible for developing and implementing strategies to reduce the company's carbon footprint, increase energy efficiency, and promote the use of renewable energy sources.

The implementation of sustainability practices involves a combination of energy efficiency measures, renewable energy adoption, and waste reduction initiatives. Companies may invest in energy-efficient equipment, optimize their supply chains, and implement recycling programs to reduce their environmental impact and improve their bottom line. Renewable energy solutions, such as solar and wind power, are also becoming increasingly important. Integrating renewable energy sources into operations helps companies reduce their reliance on fossil fuels and decrease greenhouse gas emissions, which aligns with global sustainability goals and provides long-term cost savings.

Promoting sustainable practices across the organization involves engaging employees and stakeholders in the sustainability journey. Training programs, awareness campaigns, and incentive structures can encourage employees to adopt sustainable behaviors in their daily work. The integration of digital technologies and sustainability practices requires a holistic approach. Organizations must ensure that their workforce is equipped with the necessary skills and knowledge to navigate the complexities of the ET era. This involves continuous learning, cross-functional collaboration, and a commitment to innovation. To build a skilled workforce, companies should invest in training and development programs that cover both digital technologies and sustainability practices. These programs can include workshops, online courses, and hands-on training sessions that provide employees with practical skills and knowledge.

New job roles

The ET era is driving the creation of new job roles that reflect the evolving demands of the industry. As highlighted in previous sections, the integration of digital technologies and sustainability practices necessitates a workforce with specialized skills. This section explores the emerging roles amidst the significant changes including the role of Prompt Engineers, Digital and Automation Engineers, Futurologists, Data Scientists, Sustainability Officers, and Low-Code Power App and Power BI Developers, emphasizing their importance in the ET landscape.

Prompt engineers

As AI tools become more sophisticated, roles, such as Prompt Engineers, who specialize in designing and optimizing AI prompts, are emerging.

These professionals ensure that AI systems provide accurate and actionable insights, enhancing decision-making processes. For instance, prompt engineers at energy companies work closely with AI specialists to fine-tune algorithms and models, ensuring that AI applications in predictive maintenance or energy optimization are effective and reliable. Their responsibilities include developing machine learning models, fine-tuning AI algorithms, and creating user-friendly interfaces for AI applications. This enables employees to interact seamlessly with AI tools, fostering a culture of data-driven decision-making. Their impact is profound, as their work directly contributes to optimizing operations and achieving sustainability goals.

Prompt Engineers are instrumental in bridging the gap between complex AI technologies and everyday users. They work on refining AI system interactions to be as intuitive and effective as possible. This involves understanding the nuances of how users interact with AI and continuously improving the system's responses to ensure relevance and usefulness. They are also responsible for maintaining the ethical use of AI by ensuring that the algorithms operate within the set guidelines and standards, thus promoting trust in AI systems.

Prompt engineers might develop AI systems that predict equipment failures before they occur. They would be responsible for training the AI model with historical maintenance data, refining its predictive capabilities, and ensuring the system is easy for maintenance teams to use. In this role, prompt engineers might also conduct regular training sessions for staff to ensure they are comfortable using the new system and can effectively interpret the insights generated by the AI.

AI specialists

AI Specialists focus on creating and optimizing AI solutions for various applications within the energy sector. Their work involves designing AI algorithms, developing machine learning models, and integrating AI systems into existing workflows. They collaborate with other departments to ensure that AI applications meet the specific needs of the organization, enhancing productivity, and innovation.

AI Specialists are at the forefront of technological innovation, being tasked with keeping up with the latest advancements in AI technology and determining how these can be applied to solve industry-specific challenges. This includes everything from enhancing operational efficiency to improving safety measures. Their role is highly collaborative, often requiring them to work with data scientists, engineers, and other stakeholders to develop and implement AI solutions that drive the company forward. For example, an AI specialist might develop an AI-driven platform to optimize energy storage and distribution, by analyzing patterns in energy production and consumption, and developing an AI system that can predict when to store energy and when to release it, ensuring a stable and efficient energy supply. This involves creating models that account for variables such as weather patterns, peak usage times, and maintenance schedules. The AI specialist would also be responsible for continually updating the system to improve its accuracy and effectiveness.

Digital and automation engineers

Digital and Automation Engineers play a critical role in digitizing operations. They design and implement automation systems that streamline processes, reduce human error, and increase efficiency. For example, automation engineers at ExxonMobil are developing robotic systems that automate routine inspections and maintenance tasks, freeing up human workers for more complex and strategic activities. Their responsibilities encompass a wide range of activities, from developing automation strategies to implementing and maintaining automated systems. Integrating advanced sensors and AI enables these engineers to enhance operational reliability and reduce downtime, contributing significantly to operational efficiency and cost reduction.

Low-Code Power App and Power BI developers

The rise of low-code platforms like Microsoft Power Apps and Power BI has led to the emergence of specialized developers who create custom applications and data visualizations with minimal coding. These developers enable energy companies to quickly deploy solutions tailored to their specific needs, streamlining operations, enhancing data accessibility, and driving efficiency across the organization. Leveraging these platforms allows Low-Code Power App Developers to ensure that technological advancements are accessible and usable by nontechnical staff, thereby increasing overall productivity and innovation within the company.

Low-code developers play a crucial role in creating applications that solve specific business problems. They work closely with stakeholders to understand their needs and translate them into functional applications. For instance, they might develop a workflow automation app to streamline approval processes, reducing manual tasks and speeding up decision-making.

Power BI Developers specialize in creating interactive and visually appealing dashboards and reports, transforming raw data into meaningful insights that can be easily interpreted by business users. These visualizations help in tracking Key Performance Indicators (KPIs), monitoring operational metrics, and making data-driven decisions.

Integration with existing systems is another critical aspect of their role. Developers ensure that new applications and dashboards integrate seamlessly with existing IT infrastructure. This includes connecting to various data sources, such as ERP systems, CRM software, and IoT devices, to provide a unified view of operations. For example, a Power BI dashboard might pull data from multiple sources to provide a holistic view of production efficiency. Additionally, these developers often provide training and support to ensure that users can effectively utilize the applications and dashboards. They would certainly play a leading role in an Organization's Digital Citizen program if one was setup and might conduct workshops, create user manuals, and offer ongoing support to address any issues that arise, maximizing the adoption and effectiveness of new digital tools.

Low-code developers are also responsible for maintaining and updating the applications and dashboards to ensure they remain relevant and effective. They gather user feedback, monitor performance, and implement enhancements based on evolving business needs. For instance, they might add new features to a Power Apps application based on user feedback to improve its functionality.

For example, Power Apps developer might create an App that allows technicians to log maintenance activities and issues from their mobile devices. These data can be instantly uploaded and analyzed, leading to faster resolution of problems and better tracking of maintenance trends. The app could include features, such as barcode scanning for equipment, GPS tracking for locating assets, and offline functionality, to ensure data can be captured even in remote locations. Enabling rapid development and deployment of tailored solutions, Low-Code Power App and Power BI Developers play a critical role in enhancing operational efficiency, fostering innovation, and supporting the digital transformation of energy companies. Their work ensures that employees at all levels can harness the power of digital tools to improve their workflows, make informed decisions, and contribute to the company's sustainability goals.

Futurologist

In the rapidly changing industry, having a forward-looking perspective is invaluable. Futurologists, or foresight analysts, play a critical role in energy companies by analyzing trends and predicting future developments. They help companies anticipate changes and adapt their strategies accordingly, ensuring they remain competitive and innovative in the face of evolving industry landscapes. Their insights into technological advancements and market shifts enable companies to stay ahead of the curve and make informed decisions about future investments and innovations. Engaging in extensive research and scenario planning, futurologists identify potential future developments that could impact the energy sector. This involves analyzing a wide array of data sources, including technological breakthroughs, regulatory changes, economic trends, and societal shifts. Synthesizing this information enables organizations to create forecasts to help understand potential future scenarios and prepare strategic responses. Their work is crucial for long-term planning and risk management, providing a proactive approach to navigating the uncertainties of the ET era.

For example, futurologists at International Energy Companies (IECs) might analyze emerging trends in renewable energy technologies and forecast their potential impact on the company's long-term strategy. They might examine advancements in solar and wind technologies, assess the feasibility of integrating these technologies into existing energy systems, and predict their economic viability over the next decade. Their analysis could include evaluating the potential for breakthroughs in energy storage technologies, which are critical for the widespread adoption of renewable energy. Futurologists can help senior management make informed decisions about where to allocate resources and how to position the company for future success by providing strategic recommendations.

Futurologists systematically track and analyze emerging trends in technology, economics, politics, and society, using tools and methodologies such as trend mapping, horizon scanning, and environmental scanning to identify early signs of change that could impact the energy sector. They develop multiple scenarios based on their trend analysis, outlining various potential futures and creating detailed narratives that describe how different factors could interact and evolve over time. These scenarios help companies visualize and prepare for a range of possible outcomes, enhancing their strategic agility. Working closely with senior management, futurologists integrate foresight into the company's strategic planning processes, providing insights and recommendations that inform long-term strategies, investment decisions, and innovation initiatives, which ensures that the company is well-prepared to capitalize on emerging opportunities and mitigate potential risks.

Futurologists engage with a wide range of stakeholders, including industry experts, policymakers, researchers, and customers, to gather diverse perspectives and validate their forecasts. This collaborative approach ensures that their analyses are robust and grounded in reality. They translate complex data and scenarios into clear, actionable insights, creating detailed reports, presentations, and visualizations that communicate their findings to various stakeholders, from senior executives to operational teams. Effective communication is essential for ensuring that their insights are understood and acted upon.

For instance, a futurologist might conduct a study on the potential of hydrogen as a future energy source. This study would involve analyzing technological developments in hydrogen production and storage, assessing regulatory changes that could impact the hydrogen economy, and evaluating market dynamics, such as supply and demand trends. Considering factors, such as the cost trajectory of hydrogen technologies and potential environmental benefits, supports futurologists to provide strategic recommendations to senior management. These recommendations might include investing in hydrogen research and development, forming partnerships with key players in the hydrogen value chain, and exploring pilot projects to test hydrogen applications in real-world scenarios.

Data scientists

Data Scientists are at the forefront of the energy sector's transition to a data-driven approach, playing a crucial role in leveraging vast amounts of data to drive efficiency and innovation. They are involved with analyzing extensive datasets and developing predictive models and implement machine learning algorithms, among many other tasks, all of which are integral to optimizing operations and enhancing decision-making processes. In IECs, data scientists may use advanced analytics to support optimized drilling operations, predict equipment failures, and enhance safety protocols. Their work enables energy companies to make informed decisions based on real-time data, reducing risks and improving operational performance.

Employing sophisticated statistical techniques and machine learning algorithms are an enabler for data scientists to uncover patterns and insights from historical and real-time data. This allows for the creation of predictive models that can forecast future events and trends, enabling proactive measures rather than reactive ones. For example, a data scientist might develop a predictive model to optimize drilling operations by analyzing data from previous drilling projects. This model could predict the best drilling parameters to maximize efficiency and minimize risks, leading to significant cost savings and improved safety.

Data scientists also play a critical role in enhancing asset reliability and maintenance. Through predictive maintenance models, they can analyze equipment data to forecast potential failures before they occur. This predictive capability allows for timely maintenance interventions, reducing unplanned downtime and extending the lifespan of critical assets. For instance, by analyzing vibration data, temperature readings, and other sensor data from equipment, data scientists can identify patterns that precede equipment failure, enabling preemptive maintenance actions.

In addition to operational optimization, data scientists can be utilized to enhance safety protocols by analyzing data from various sources, such as incident reports, sensor data, and environmental conditions and developing models that predict safety risks and help in implementing preventive measures that protect workers and assets.

Data scientists can also collaborate with other departments to integrate data-driven insights into the overall business strategy. They work closely with engineers, operations teams, and management to ensure that their models and recommendations align with the company's goals and are practically implementable. This collaboration ensures that data-driven strategies are effectively translated into actionable plans that enhance the company's performance.

Sustainability Officers

With an increasing focus on sustainability, the role of Sustainability Officers has become more prominent in the energy sector. These professionals are tasked with developing and implementing strategies to reduce the environmental impact of energy operations. Their work spans various initiatives, including carbon reduction, renewable energy integration, and ensuring compliance with environmental regulations. Sustainability Officers play a crucial role in aligning company operations with global sustainability goals and regulatory requirements, driving both environmental and business performance.

Sustainability Officers are responsible for creating comprehensive sustainability plans that outline the company's goals, strategies, and actions for reducing its environmental footprint. This involves conducting thorough assessments of current operations to identify areas where improvements can be made. They often work on projects that aim to decrease greenhouse gas emissions, enhance energy efficiency, and increase the use of renewable energy sources.

In addition to technical projects, Sustainability Officers engage in policy development and advocacy. They monitor evolving environmental regulations and ensure the company remains compliant with all relevant laws. This includes preparing and submitting detailed reports to regulatory bodies, participating in environmental audits, and advocating for policies that support sustainable practices within the industry. Their role is critical in maintaining the company's reputation and avoiding legal penalties associated with noncompliance.

Sustainability Officers also focus on integrating sustainability into the corporate culture. They work to raise awareness among employees about the importance of sustainable practices and provide training on how to implement these practices in their daily work. This involves organizing workshops, creating educational materials, and leading corporate social responsibility initiatives. These officers ensure that environmental considerations are embedded in every aspect of the company's operations.

Collaboration is a key aspect of the Sustainability Officer's role. They work closely with other departments, such as operations, engineering, and finance, to ensure that sustainability goals are integrated into broader business strategies. This interdisciplinary approach ensures that sustainability initiatives are practical, cost-effective, and aligned with the company's overall objectives. They also engage with external stakeholders, including government agencies, nonprofit organizations, and the community, to build partnerships and support for sustainability projects.

Cybersecurity Analyst

As digital technologies become more integral to energy operations, the role of Cybersecurity Analysts has become increasingly critical. These professionals are responsible for safeguarding critical infrastructure from the ever-evolving landscape of cyber threats. The energy sector, with its extensive and interconnected digital systems, faces unique cybersecurity challenges. A cyberattack on an energy company can lead to severe disruptions, financial losses, and even safety hazards. Therefore, Cybersecurity Analysts play a pivotal role in maintaining the integrity and security of digital systems.

Cybersecurity Analysts monitor networks for suspicious activity, utilizing advanced tools and technologies to detect potential threats in real-time. This involves analyzing vast amounts of data to identify anomalies and patterns that may indicate a cyberattack. In addition to monitoring, Cybersecurity Analysts are tasked with implementing robust security protocols to protect against threats. This includes setting up firewalls, encryption mechanisms, and access controls to secure sensitive data and systems. They ensure that only authorized personnel have access to critical infrastructure, reducing the risk of insider threats. Regularly updating and patching software to fix vulnerabilities is also a key part of their responsibilities, as outdated systems can be easily exploited by attackers.

Responding to security breaches is another crucial aspect of the Cybersecurity Analyst's role. When a breach occurs, they must act swiftly to contain the threat, minimize damage, and restore normal operations. This involves conducting forensic investigations to determine the source and extent of the breach, eradicating the threat, and implementing measures to prevent future incidents. Effective incident response plans are essential for minimizing the impact of cyberattacks and ensuring a rapid recovery. Cybersecurity Analysts also play a proactive role in threat intelligence and risk assessment. They stay abreast of the latest cybersecurity trends, emerging threats, and vulnerabilities by participating in industry forums, collaborating with other organizations, and conducting ongoing research.

The emergence of these new job roles highlights the transformative impact of the digital revolution on the energy sector. As companies continue to adopt digital technologies and data-driven approaches, the demand for skilled professionals in these areas will only grow. Investing in workforce development and fostering opportunities for continuous learning ensures that energy companies have the talent needed to thrive in the digital era. This strategic focus on emerging roles and skills is vital for navigating the complexities of the energy transition and achieving long-term success. These new roles are crucial in driving innovation, optimizing operations, and ensuring a sustainable and secure energy future, it is important therefore to embrace the change and where appropriate to adopt these new job roles.

Impact of new roles on energy companies

With the introduction of new jobs roles, traditional organization structures of energy companies are being reshaped. These roles drive innovation and operational efficiency, necessitating the creation of new departments or the extension of existing ones to harness their specialized functions. This strategic restructuring supports a culture of continuous improvement and adaptation in the rapidly evolving energy sector.

These new roles leverage digital technologies and data-driven approaches, leading to advancements in operational efficiency and innovation. Prompt Engineers and Data Scientists collaborate to develop sophisticated AI models that optimize maintenance schedules and energy consumption, reducing downtime and operational costs. Digital and Automation Engineers work with Sustainability Officers to implement automation solutions that are both efficient and environmentally friendly, ensuring that energy companies meet their sustainability targets. Such collaborations result in streamlined operations, enhanced productivity, and a stronger competitive edge.

To support the specialized functions of these new roles, energy companies may establish new dedicated departments or functions. Some examples are provided next, these may be stand alone or combinations depending on the specific requirements of each organization.

Digital transformation department

This department oversees the integration of digital technologies across the organization. It includes roles such as Prompt Engineers, Digital and Automation Engineers, and Low-Code Power App Developers. The department ensures a coordinated approach to digital innovation and implementation with the rest of the organizations departments and operations, focusing on transforming traditional processes through technology. The primary objectives are to enhance digital literacy, streamline operations through automation, and foster a culture of continuous improvement.

Data analytics and visualization department

A Data Analytics and Visualization Department plays a pivotal role in leveraging data to drive efficiency, innovation, and strategic decision-making. This department, staffed with data scientists and Power BI developers, enhances the company's data analysis and visualization capabilities, turning raw data into actionable insights. The department provides deep insights into various aspects of the business, including operational performance, customer behavior, and market trends by employing advanced analytical techniques and visualization tools. These insights are crucial for identifying opportunities for improvement, optimizing processes, and anticipating future challenges, thereby facilitating informed decision-making and strategic planning.

The department develops custom dashboards and analytical tools tailored to the specific needs of the organization, ensuring that data-driven insights are readily accessible to all decision-makers. These dashboards consolidate KPIs and other relevant data into intuitive, interactive visualizations that allow users to quickly grasp complex information and trends. This accessibility empowers executives, managers, and frontline employees to make informed decisions based on real-time data. Furthermore, the department collaborates closely with other business units to identify data needs, develop predictive models, and create algorithms that enhance forecasting accuracy and operational efficiency. Fostering a data-driven culture enables the Data Analytics and Visualization Department to enhance the organization's competitive edge and drive continuous improvement and innovation across all facets of the business.

AI and ML Department

The AI and ML Department is dedicated to harnessing the power of AI and ML to drive innovation, efficiency, and strategic advantage. Comprising AI Specialists, Prompt Engineers, and Data Scientists, this department focuses on developing and deploying AI solutions tailored to the unique needs of the company. Its objectives are dependent on each company's specific needs, these could include predictive maintenance, optimizing energy usage, enhancing decision-making processes, and exploring new AI technologies and their potential applications within the organization.

A practical example of how this organization will add value is in predictive maintenance. Analyzing data from equipment sensors and historical maintenance records enables AI models to predict potential failures before they occur, allowing for timely intervention and reducing downtime. This improves the reliability of critical infrastructure as well as reduces maintenance costs and extends the lifespan of equipment. The department develops sophisticated algorithms that monitor real-time data, identifying patterns and anomalies that indicate potential issues, ensuring the smooth operation of the company's assets.

The department also focuses on enhancing decision-making processes through AI-driven insights by developing and deploying AI tools that analyze market trends, customer behavior, and operational data. The outcome can provide executives and managers with actionable insights that inform strategic planning and operational decisions. These tools can process and interpret complex datasets, uncovering hidden patterns and trends that human analysis might miss. This capability enables the company to stay ahead of market changes, respond proactively to emerging challenges, and capitalize on new opportunities.

In addition to these core functions, the AI and ML Department is dedicated to researching and exploring new AI technologies. This involves developing proprietary AI tools and solutions that address specific challenges and opportunities within the energy sector. These innovations provide the company with unique operational efficiencies and capabilities that differentiate it from competitors.

Digital Twin Technology Department

A Digital Twin Technology Department can be established to create and manage digital replicas of physical assets and systems. Digital twins enable real-time monitoring, predictive maintenance, and optimization of assets, leading to increased efficiency and reduced downtime. Key functions of this department include developing digital twin models for critical infrastructure, such as power plants, pipelines, and grids, integrating IoT sensors and data analytics to provide real-time insights, and using simulations to predict and prevent potential issues. For instance, a digital twin of an offshore oil platform could simulate various scenarios, such as extreme weather conditions, to optimize safety protocols and operational efficiency.

Innovation Lab and Incubator

An Innovation Lab and Incubator serves as a hub for fostering a culture of innovation and supporting the development of groundbreaking technologies and business models, and it will work closely with the other digital departments, such as AI and Machine Learning, Data Analytics and Visualization, and Digital Transformation Departments. This initiative encourages employees to experiment with new ideas and collaborate on innovative projects. The lab's key functions include providing resources and support for pilot projects and prototypes, facilitating collaboration with startups, research institutions, and other external partners, and organizing hackathons, workshops, and innovation challenges.

In practice, the Innovation Lab offers a dedicated space equipped with the latest tools and technologies, enabling employees to turn their ideas into tangible solutions. Providing access to funding, mentorship, and technical expertise enables the lab to help transform promising concepts into viable products or processes. Collaboration is a cornerstone of the lab's activities, as it actively seeks partnerships with external entities, such as startups, academic institutions, and industry experts. These collaborations bring fresh perspectives and specialized knowledge, enhancing the lab's ability to tackle complex challenges and explore new frontiers.

   Moreover, the Innovation Lab and Incubator plays a crucial role in cultivating an innovation mindset across the organization. Through workshops and training sessions, employees are equipped with the skills and knowledge needed to think creatively and approach problems from different angles. The lab also implements a structured process for idea submission and evaluation, ensuring that the best ideas receive the attention and resources they need to flourish.

Cybersecurity department

Although cybersecurity is nothing new, there is a marked increase in threats in the digital landscape, so therefore far more focus on Cybersecurity is needed, potentially with the establishment of a dedicated department. This department includes Cybersecurity Analysts and works closely with the IT Department to protect critical infrastructure. It is responsible for monitoring networks, implementing security protocols, and responding to security breaches, ensuring the integrity and security of the company's digital assets. Proactive measures, such as regular security audits, penetration testing, and continuous monitoring, are key functions to safeguard against potential cyber threats.

Sustainability department

Traditionally, sustainability was embedded within the Health, Safety, and Environment (HSE) department. Incorporating sustainability within a dedicated department offers significant advantages for energy organizations. A focused sustainability department can provide specialized attention to environmental and social governance issues, allowing for more strategic and comprehensive approaches to sustainability. This separation ensures that sustainability receives the necessary resources and expertise, distinct from the immediate operational safety concerns typically managed by HSE. It fosters innovation, enables the setting of ambitious sustainability goals, and ensures that these objectives are integrated into the core business strategy. As sustainability challenges grow increasingly complex and critical to long-term success, having a dedicated department highlights the organization's commitment to sustainable development, enhancing its reputation and appeal among stakeholders, investors, and customers.

   A sustainability department in an energy organization plays a critical role in ensuring environmentally responsible and socially accountable operations. This department develops and implements comprehensive sustainability strategies aligned with the organization's goals, formulating policies on environmental stewardship, energy efficiency, and waste management. The department drives the company's environmental agenda by setting both short-term and long-term sustainability goals, such as reducing carbon emissions and increasing renewable energy usage. It also oversees energy-saving measures, renewable energy integration, resource optimization, and pollution control, thereby minimizing the environmental impact of the organization's activities. In addition, the department prepares sustainability reports according to international standards, ensures compliance with environmental laws, and achieves certifications like ISO 14001 and ISO 50001.

The sustainability department also engages stakeholders, including investors, customers, employees, and local communities, to address their sustainability concerns and expectations. It conducts awareness campaigns, fosters collaborations with government bodies, and promotes sustainable supply chain management through supplier audits and green procurement. Employee training and involvement are crucial, with programs designed to educate and engage the workforce in sustainability initiatives. The department also identifies and mitigates environmental and social risks, develops community-benefiting projects, ensures fair labor practices, and promotes diversity and inclusion. Through these efforts, the sustainability department not only ensures regulatory compliance but also positions the organization as a leader in sustainable practices, enhancing its reputation and long-term viability.

In the midst of the digital era and the energy transition, the rapid pace of change is creating a wealth of opportunities across the energy sector. This dynamic landscape is driving the emergence of new roles and the formation of specialized departments designed to support and leverage these opportunities. Innovations in technology and sustainability are not only reshaping traditional operations but also opening avenues for advanced analytics, AI-driven solutions, and cutting-edge research and development. As energy companies adapt to these shifts, they are establishing departments, such as AI and Machine Learning, Data Analytics and Visualization, and Innovation Labs to stay ahead of the curve. These departments foster innovation, improve operational efficiency, and enhance decision-making processes, ensuring that the organization is well-equipped to thrive in this transformative era. This proactive approach to embracing change positions energy companies to capitalize on new opportunities, drive sustainable growth, and lead the way in the evolving energy landscape.

Best practices for workforce transformation

A prominent example of workforce transformation is demonstrated by a major oil and gas company that embarked on an ambitious digital training initiative. Recognizing the critical importance of equipping their workforce with modern digital knowledge and skills, the company rolled out a comprehensive training program targeting all levels of employees. This initiative aimed to foster a digital-first mindset and ensure that employees were proficient in utilizing advanced technologies to drive the company's strategic goals.

To design the digital training program, the company adopted a blended learning approach that combined workshops, online courses, and hands-on training sessions. This multifaceted strategy was crafted to cater to different learning preferences and ensure maximum engagement and retention of new skills. The workshops were designed as interactive sessions, providing employees with real-world applications of digital tools and technologies, such as data analytics, AI, and machine learning. To accommodate the diverse schedules and locations of employees, a suite of online courses was developed, offering flexibility and allowing individuals to learn at their own pace. These courses covered a wide range of topics from basic data literacy to advanced AI algorithms. Additionally, the company emphasized the importance of practical experience through hands-on training sessions, where employees could apply their newfound knowledge to actual projects. This practical component was crucial for reinforcing learning and demonstrating the tangible benefits of digital technologies in real-world scenarios.

Workshops: These hands-on sessions focused on real-world applications of digital tools and technologies. Employees participated in interactive workshops covering topics such as data analytics, AI, and machine learning. These sessions were designed to build practical skills that could be immediately applied in their daily tasks. For instance, in one workshop, participants used Power BI to create custom dashboards that visualized operational data, enabling more informed decision-making. The workshops provided employees with the tools and knowledge necessary to interpret and present data effectively.

Online courses: To ensure flexibility and accessibility, the company developed a suite of online courses. These courses allowed employees to learn at their own pace, providing modules on various digital tools, software, and emerging technologies. Topics ranged from basic data literacy to advanced AI algorithms. This approach ensured that all employees, regardless of their location or schedule, had the opportunity to upskill. For example, a course on AI-driven predictive maintenance taught employees how to leverage machine learning models to anticipate equipment failures and plan maintenance activities proactively. This training directly supported the company's efforts in adopting advanced predictive maintenance practices.

Hands-on training sessions: In addition to theoretical knowledge, the company emphasized the importance of practical experience. Employees engaged in hands-on training sessions where they could apply their newly acquired skills to real projects. These sessions were crucial for reinforcing learning and demonstrating the tangible benefits of digital technologies. For instance, in a hands-on session focused on energy optimization, teams worked on live data from the company's operations to identify inefficiencies and propose data-driven solutions. These practical experiences were instrumental in demonstrating how digital tools could be applied to improve operational efficiency.

Implementing a comprehensive digital training program has brought transformative benefits to the energy company, positioning it at the forefront of industry innovation. One of the key advantages is the significant enhancement in operational efficiency. Employees equipped with up-to-date digital skills are empowered to utilize data analytics effectively, streamline processes, and identify and resolve bottlenecks promptly. Real-time insights provided by custom dashboards have allowed for more informed decision-making, further contributing to efficiency gains. These tools enable employees to visualize data trends and operational metrics clearly, which helps in pinpointing areas that need improvement and implementing targeted solutions quickly. This improved operational performance has reduced costs and enhanced the company's overall productivity in certain areas. Additionally, the ability to interpret and act on data swiftly ensures that the company can maintain high standards of service delivery and operational excellence, even as it scales its operations.

Another major benefit of the digital training program is the substantial reduction in equipment downtime and the overall uplift in performance. Employees, now adept in predictive maintenance techniques, can analyze data patterns to anticipate and address issues before they lead to costly shutdowns. This proactive approach to maintenance has been made possible through the use of AI models that predict potential equipment failures based on historical and real-time data. As a result, maintenance teams can schedule interventions at the most opportune times, thereby avoiding unplanned outages and extending the lifespan of critical machinery.

Continuously focusing on innovation and improvement and investing in digital upskilling has enabled the company to develop a digitally empowered workforce capable of driving sustainable growth. This ongoing commitment to digital skills development ensures that the workforce remains adept at using the latest technologies, thereby future proofing the organization against industry disruptions and positioning it for long-term success.

Renewable energy company: Integration of new job roles

Another illuminating case study involves a leading renewable energy company that embarked on a transformative journey by embracing the creation and integration of new job roles specifically tailored for the digital and energy transition era. This forward-thinking company recognized the vast potential of AI and automation to revolutionize its operational efficiency and sustainability efforts. With the dual objectives of enhancing technological capabilities and driving environmental stewardship, the company initiated a strategic overhaul of its workforce and operational practices.

This comprehensive transformation began with a detailed analysis of the existing workforce capabilities and future needs aligned with the company's strategic vision. Recognizing that traditional roles and skill sets were insufficient to harness the full potential of emerging technologies, the company prioritized the integration of specialized roles. By doing so, they aimed to infuse the organization with cutting-edge expertise in AI, ML, and automation, which were identified as critical drivers of future growth and sustainability.

The company meticulously designed new roles, such as Digital and Automation Engineers, focusing on digitizing and automating core processes. This strategic addition of new job functions was intended to drive a fundamental shift in how the company approached its operations. The introduction of these roles was seen as essential for building an agile and technologically adept workforce capable of leading the company through the digital transformation and the ongoing energy transition.

In parallel, the company revamped its operational practices to align with the capabilities brought by these new roles. This involved integrating AI and automation into everyday workflows. The company also implemented comprehensive training and onboarding programs to ensure that new hires and existing employees could collaborate effectively. These programs included regular training sessions and workshops to upskill current employees, enabling them to work efficiently alongside AI specialists and automation engineers. This inclusive strategy fostered a culture of continuous learning and adaptation, crucial for the smooth adoption of new technologies and practices.

The long-term benefits of integrating new job roles and embracing digital transformation are profound. The company achieved notable improvements in operational efficiency and cost savings. Automation of routine tasks allowed employees to focus on more strategic activities, enhancing overall productivity. The precision and efficiency of AI-driven processes optimized resource usage, reduced waste, and lowered operational costs.

Furthermore, the emphasis on sustainability and innovation significantly enhanced the company's brand image, attracting environmentally conscious investors and customers. This alignment with global sustainability goals opened up new business opportunities, such as partnerships and funding for green projects, further driving growth.

The strategic integration of new job roles, departments, and digital technologies serves as a model for other organizations seeking to navigate the complexities of the digital and energy transition eras successfully.

Best practices

The success stories of companies described previously offer valuable lessons and best practices for others in the energy sector looking to navigate workforce transformation effectively. These strategies can support a smooth transition into the digital and energy transition era but also position companies for long-term success and sustainability.

1. Invest in continuous learning

First and foremost, developing a culture of continuous learning is essential for any organization aiming to stay ahead in the rapidly evolving energy sector. Companies should invest in comprehensive training programs that keep employees updated with the latest digital and technological advancements. This approach can enhance knowledge and skills as well as instill a mindset of innovation and adaptability. A good example is the comprehensive digital training program described in the previous example, which included a blend of workshops, online courses, and hands-on training sessions aimed at different levels in the organization to equip employees with skills in data analytics, AI, and ML. Such initiatives ensure that employees are firstly aware of these new digital tools that can vastly improve productivity and efficiency in their jobs and also so they can effectively utilize these new tools and technologies, which helps to drive the efficiency and innovation across the entire organization.

Continuous learning programs should be tailored to address specific skill gaps identified through assessments of workforce capabilities. These programs can include certifications, advanced courses, and partnerships with educational institutions to provide employees with cutting-edge knowledge. Additionally, fostering a learning environment where employees are encouraged to pursue new skills and knowledge can lead to greater job satisfaction and retention, further contributing to the company's success.

2. Embrace new job roles

Identifying and accepting that there is a need for new job roles to address current organizational gaps to meet the specific demands of the energy transition era is crucial. Roles, such as Prompt Engineers, Digital and Automation Engineers, and Futurologists, are designed to drive technological adoption and strategic foresight. The integration of Prompt Engineers for example enables the development of AI-driven solutions that streamlined workflows and enhanced decision-making processes. Introducing these specialized roles enables companies to build a workforce capable of leading digital transformation initiatives and driving sustainability efforts. Moreover, these new roles should be supported by clear career pathways and development opportunities. Providing employees with a vision of how they can grow within the company and the impact they can have on its strategic goals can motivate them to embrace new technologies and approaches. Additionally, fostering a culture that values and rewards innovation can further encourage employees to take on these new roles and responsibilities.

3. Leverage data and AI

Utilizing data and AI tools can significantly improve operational efficiency and decision-making. Companies should focus on building data literacy across the organization and integrating AI-driven solutions into their workflows. In the example previously, the use of AI in predictive maintenance for example allows the companies to foresee equipment failures and schedule timely interventions, reducing downtime and maintenance costs. By leveraging data analytics and AI, companies can optimize their operations, reduce costs, and make more informed strategic decisions.

To effectively leverage data and AI, companies should invest in robust data infrastructure and analytics platforms. This includes implementing tools for data collection, storage, and analysis, as well as training employees to interpret and act on data insights. Additionally, fostering a data-driven culture where decisions are based on empirical evidence rather than intuition can lead to more accurate and efficient outcomes.

4. Foster collaboration

Encouraging collaboration between new and existing roles can create synergies that enhance overall performance. Cross-functional teams can drive innovation and ensure that various aspects of digital transformation and sustainability are addressed cohesively. Integrating new roles involved cross-functional teams can encourage collaboration between existing staff and newly hired experts, which can in turn improve operational efficiency and help to instill a culture of continuous learning.

Creating platforms and opportunities for collaboration, such as innovation labs, workshops and even regular team meetings, can help break down silos and encourage knowledge sharing. Additionally, promoting a culture of inclusivity where diverse perspectives are valued can lead to more creative and effective solutions. Ensuring that collaboration is supported by the right tools and technologies, such as communication platforms and project management software, can also enhance the effectiveness of cross-functional teams.

5. Prioritize sustainability

Incorporating sustainability into core operations is not just a regulatory requirement but a strategic imperative. Roles dedicated to sustainability can help companies achieve their environmental goals and enhance their corporate reputation. By integrating sustainability into business strategies, companies can attract environmentally conscious investors and customers, opening up new business opportunities and enhancing their market position.

To prioritize sustainability, companies should set clear environmental goals and metrics to track progress. This includes implementing green initiatives, such as renewable energy projects, energy efficiency programs, and waste reduction strategies. Additionally, engaging employees in sustainability efforts through awareness programs and incentives can foster a culture of environmental responsibility. Collaborating with external partners, such as research institutions, can also enhance a company's sustainability initiatives by bringing in additional expertise and resources.

Clearly, there are many more best practices; however, these key ones help to highlight the transformative potential of a well-prepared workforce in the ET. Investing in continuous learning, embracing new job roles, leveraging data and AI, fostering collaboration, and prioritizing sustainability empowers energy companies to navigate the complexities of the energy transition and digital era. The success stories of companies described previously serve as valuable examples for others in the sector, demonstrating how strategic workforce transformation can drive operational excellence, cost efficiency, and environmental stewardship.

Embracing leadership for the new energy workforce

As the energy sector stands on the brink of a monumental transformation, the shift toward renewable energy sources is fundamentally reshaping the landscape of energy production, distribution, and consumption. As we navigate this transition, we witness profound changes across the entire energy value chain. The energy mix is evolving rapidly, and the digital revolution is playing a pivotal role. Although this transition will not happen overnight, fossil fuels will continue to play a role in the foreseeable future, but the trajectory is clear: the energy sector is undergoing an irreversible transformation. New opportunities are emerging for companies ready to adapt, innovate, and lead. Conversely, those that fail to embrace these changes or adapt too slowly will face significant challenges and potential obsolescence.

In this rapidly evolving landscape, embracing change, agility, and proactivity are essential. Companies must be willing to reinvent themselves, acquire new skills, and forge partnerships within entirely new ecosystems. Successfully navigating this transformation requires a robust embrace of innovation and a keen pursuit of new opportunities. This may involve exploring novel business models, investing in groundbreaking technologies, and adopting fresh approaches to collaboration and partnerships.

Ultimately, the goal is to create a more sustainable, efficient, and technologically advanced energy industry capable of meeting the demands of future generations. We have discussed the need for the critical role of the workforce in this transition, examining the skills, roles, and cultural shifts needed to thrive in the new landscape. In particular, we have explored how companies can leverage digital advancements and data-driven strategies to empower their employees and drive sustainable growth.

The next challenge is leadership. Leading this new workforce requires a distinct type of leadership, one that is forward-thinking, agile, and deeply committed to innovation and sustainability. Leaders must foster a culture of continuous learning and adaptability, ensuring that their teams are equipped with the latest skills and technologies. They must champion the integration of digital tools and data-driven strategies, promoting a mindset that values experimentation and rewards innovation.

Moreover, new leaders must prioritize sustainability, embedding it into the core of their business strategies. This involves setting clear environmental goals, engaging stakeholders in sustainability efforts, and ensuring compliance with regulatory requirements. By doing so, leaders can drive both environmental and business performance, attracting environmentally conscious investors and customers.

The energy transition is not just about adopting new technologies but about fundamentally rethinking how the workforce operates and how leadership can guide this transformation. Embracing these changes and investing in the development of a skilled, adaptable workforce will be crucial for navigating the complexities of the energy transition and achieving long-term success in a rapidly evolving industry. The next chapter will further explore the leadership qualities and strategies required to lead this new workforce effectively.

References

[1] Shell, C3.ai, Baker Hughes, and Microsoft, Shell, C3 AI, Baker Hughes, and Microsoft Launch the Open AI Energy Initiative, an Ecosystem of AI Solutions to Help Transform the Energy Industry, 2021.
[2] H. Mason, Data-Driven: Creating a Data Culture, O'Reilly Media, 2015.

CHAPTER TEN

A new age of leadership

As the global energy sector undergoes a monumental transformation, the traditional leadership styles that have dominated for decades are increasingly being called into question. Are traditional CEOs about to become a vestige of the past? With the rapid evolution of the energy landscape, it is becoming clear that the leadership styles that served us in the past may no longer be appropriate for the future. Historically, many CEOs prioritized numbers and bottom-line impact as the singular metric of performance, often resorting to downsizing and cutting human labor costs to rebalance the books. However, this approach is increasingly seen as outdated in an era where innovation, sustainability, and agility are paramount.

As companies reinvent themselves and develop new Energy Transition (ET) projects while pivoting toward greener portfolios, new challenges and uncertainties have become the norm. Traditional contracting approaches have proven cumbersome, necessitating a shift toward more progressive and flexible models as described in Chapter 3. To navigate this complex environment, a new type of leadership is required one that can think strategically, break the mold, and make courageous decisions regarding project delivery models.

This new approach to contracting and project management demands leaders who prioritize collaboration, communication, and innovation. Leaders must be willing to invest in new technologies and solutions that support these values. They must also be prepared to overcome cultural and organizational barriers that may prevent effective collaboration with partners. Embracing a more collaborative mindset is essential, as the energy sector faces multifaceted challenges that cannot be addressed by isolated efforts.

The COVID-19 pandemic has left an indelible mark on the world of work, accelerating the acceptance of remote working and digital collaboration. This shift has further emphasized the need for adaptable leadership. Leaders now must manage teams that are often geographically dispersed, leveraging digital tools to maintain communication and productivity. The ability to foster a cohesive and motivated team in a remote work environment is becoming a critical skill for modern executives.

Powering Through the Transition
https://doi.org/10.1016/B978-0-323-91754-4.00011-X

New leaders must be able to think creatively and take bold action. The sector is facing significant challenges, but it also holds enormous potential for innovation and growth. These leaders must be willing to take risks and try new things, even in the face of uncertainty. This might include investing in new technologies or business models that are not yet proven, or pursuing partnerships or collaborations with companies or organizations outside the traditional energy sector.

Leaders must also be prepared to overcome cultural and organizational barriers that hinder effective partnerships. Embracing a mindset of continuous improvement and openness to change is important for new age leaders so they can position their organizations to thrive in this dynamic environment.

The energy sector's transition is not just about adopting new technologies but fundamentally rethinking how companies operate and collaborate. This requires leaders who can inspire and guide their teams through periods of significant change, encouraging innovation and resilience. The future of energy leadership lies in the ability to adapt, innovate, and lead with a vision that embraces both the challenges and opportunities of the energy transition and digital era.

The shifting leadership paradigm in the energy sector demands a new breed of leaders who are not only strategically astute but also courageous, collaborative, and innovative. As we move forward, these leaders will play a pivotal role in driving the transformation toward a more sustainable and technologically advanced energy industry. This chapter will explore the essential qualities of this new age of leadership.

The changing landscape of leadership

Traditional leadership models are often characterized by hierarchical structures and rigid strategies. Historically, these models focused heavily on top-down decision-making, stability, and predictability. Autocratic and bureaucratic styles of leadership prioritized control and efficiency, relying on clear chains of command and a centralization of authority. These models were effective in an era where stability and incremental improvements were the norm. However, they are ill-suited to handle the fast-paced changes, challenges, and complexities of the ET and digital era.

One of the major limitations of traditional leadership is its hierarchical nature, which stifles innovation and responsiveness. Decisions made at the top often lack the nuanced understanding of frontline challenges, leading to slow responses to market changes and technological advancements. For example, in the past, large energy companies operated under highly centralized decision-making frameworks. While this approach ensured consistency and control, it also meant that the company was slow to react to the rising importance of renewable energy sources and digital technologies. This lag in adaptation may have resulted in missed opportunities and the need for costly strategic pivots later on.

In traditional leadership models, control is emphasized over collaboration. This approach can be detrimental in today's interconnected world. The energy sector's integration of new technologies, such as artificial intelligence (AI) and automation, requires a collaborative approach, bringing together diverse teams of engineers, data scientists, and operational staff. Traditional leadership styles, with their siloed structures, often hinder this necessary cross-functional collaboration.

Moreover, traditional leadership models typically focus on short-term results and efficiency metrics. While these are important, they often lead to a reactive rather than proactive approach. In the context of the energy transition, leaders must balance short-term operational goals with long-term sustainability objectives. Traditional leadership's focus on immediate gains can undermine efforts to invest in future technologies and sustainable practices. For instance, many traditional energy companies have historically focused on maximizing short-term profits from fossil fuels, delaying investments in renewable energy. This short-termism has left them vulnerable to market shifts toward greener energy solutions.

Traditional leadership models also tend to resist change, clinging to established practices and technologies even when they become obsolete. This resistance can be particularly problematic in the fast-evolving energy sector, where adaptability is crucial. Leaders who rely on traditional methods may find themselves unprepared to handle the rapid advancements in technology and shifts in market dynamics. An example of this is the slow adoption of digital tools and renewable energy technologies by some oil and gas companies, which has placed them at a competitive disadvantage compared to more innovative peers [1].

Furthermore, the rigid strategies associated with traditional leadership often lack the flexibility needed to navigate the uncertainties of the energy transition and digital era. These strategies are typically designed for predictable environments, making it difficult for organizations to pivot quickly in response to unexpected challenges or opportunities. This inflexibility can hinder a company's ability to explore new business models or adopt emerging technologies, both of which are essential for thriving in the current landscape.

The limitations of traditional leadership models are increasingly evident in the face of the energy transition and the digital era. Hierarchical structures, a focus on control, and short-termism are ill-suited to the fast-paced and complex challenges of today. Leaders in the energy sector must recognize these shortcomings and prepare to adopt more dynamic and flexible leadership styles to navigate the transformative changes ahead effectively.

Outdated traditional methods: The case of a multinational oil and gas conglomerate

A compelling example of how traditional leadership methods can hinder a company's ability to adapt and thrive in the fast-evolving energy sector. A major multinational conglomerate, which operated part of its business in the Oil and Gas Sector, struggled with outdated leadership approaches that impacted its competitiveness in the rapidly changing energy landscape. It operated under a traditional hierarchical structure with decision-making concentrated at the top levels of management.

This hierarchical structure led to several key issues, firstly the company exhibited a slow response to market changes. Due to lengthy approval processes required for strategic decisions, the company often lagged in adapting to new market conditions. For instance, as the demand for renewable energy technologies grew, competitors swiftly pivoted to offer new solutions, which left this company behind with its bureaucratic structure causing delays in entering the renewable energy market, resulting in a significant loss of market share.

Secondly, the top-down decision-making approach stifled innovation within the company. Lower-level managers and frontline employees, who were more attuned to operational challenges and market opportunities, had limited input in strategic decisions. This restriction meant that innovative ideas from these employees rarely reached senior management, leading to missed opportunities for technological advancements and process improvements. The lack of a platform for these employees to contribute ideas hindered the company's ability to innovate and stay competitive in a rapidly evolving industry.

Control over collaboration

Traditional leadership styles often emphasize control and adherence to established procedures, which can lead to several problems within a company. One major issue is the inadvertent introduction of siloed operations where departments tend to operate independently with minimal cross-functional collaboration. For example, for this company, engineering teams may work separately from data analytics teams, missing opportunities to integrate advanced data analytics into engineering solutions. This siloed approach inhibits the company from developing integrated, innovative solutions for its clients, ultimately limiting its competitive edge.

Another key issue was resistance to new technologies. Senior management in this company, were frequently showing reluctance to adopt new technology. This resistance is particularly evident in the slow adoption of digital tools. While competitors leverage these technologies to enhance their service offerings and operational efficiency, companies following traditional methods continue to rely on outdated systems and practices. This reliance on traditional methods hampers their ability to innovate and also puts them at a significant disadvantage in the rapidly evolving market. Consequently, this company found itself lagging behind more agile and innovative competitors in the market.

The traditional emphasis on control and established procedures within leadership can significantly hinder a company's ability to innovate and adapt to new technologies. Siloed operations and resistance to technological advancements are clear indicators of these limitations.

Struggles and reorganization

These challenges extended beyond the company's oil and gas division, affecting the broader company and leading to significant financial difficulties. In response to years of declining performance and strategic missteps, a comprehensive reorganization and strategic shift resulted. The reorganization included significant layoffs and divestitures, focusing on core businesses while shedding noncore assets. This strategic shift was necessary to streamline operations and focus on areas with the highest potential for growth and innovation. However, the process also revealed the deep-rooted issues within the company's traditional leadership model.

This reorganization aimed to address the overreliance on traditional business models and the company's failure to adapt quickly to market changes. Its hierarchical, top-down structure played a crucial role in its inability to innovate rapidly and respond to new market demands, such as the increasing shift toward renewable energy. This rigid structure hindered the flow of innovative ideas and slowed decision-making processes, preventing the company from keeping pace with more agile competitors.

The case of this multinational conglomerate highlights the limitations of traditional leadership models in the modern energy sector. Hierarchical structures that emphasize control and efficiency over flexibility and innovation are ill-suited to handle the complexities of the energy transition. The importance of evolving leadership styles to embrace agility, innovation, and sustainability cannot be underestimated. The limitations of traditional leadership models, with their focus on control, hierarchical decision-making, and short-termism, are starkly evident. To navigate the complexities of the energy transition and digital era successfully, energy companies must shift toward more agile, innovative, and sustainable leadership approaches. Embracing these new leadership paradigms is essential for fostering innovation, enhancing collaboration, and ensuring long-term strategic success.

Agile leadership, therefore, emerges as a vital paradigm for the ET and digital era. Agile leadership prioritizes adaptability, empowering leaders to pivot strategies in response to new information and emerging trends. This approach is particularly relevant in the context of the digital revolution, where advancements in AI, Machine Learning (ML), and data analytics are continuously reshaping industry operations. As highlighted in previous chapters, the successful implementation of these technologies hinges on leaders who can guide their teams through ongoing changes, fostering a culture of learning and experimentation.

Agile leadership a vital paradigm

Building on the conconcept of Agile Management presented in Chapter 8, agility in leadership refers to the ability to move quickly and easily, adapting strategies and operations to meet changing circumstances. It involves being responsive to new information, technologies, and market conditions, making decisions that align with evolving goals. Agile leaders are characterized by their openness to innovation, willingness to take calculated risks, and ability to pivot when necessary. This agility allows leaders to stay ahead of the curve, continuously adapting to the fast-paced changes that define the energy transition and digital era.

Agile leadership is the ability to lead organizations through rapid changes and uncertainties by adopting flexible, collaborative, and adaptive strategies. This style of leadership is particularly relevant in today's fast-paced and ever-evolving business environment. Leaders who embrace agility are able to quickly pivot and adjust their strategies in response to new information, technologies, and market conditions. They foster a culture of innovation and continuous improvement, empowering their teams to experiment, learn from failures, and iterate on solutions.

In the context of the energy transition, agile leadership becomes crucial due to the fast-paced and dynamic nature of changes in the energy sector. The energy industry is undergoing a monumental shift, moving away from traditional fossil fuels toward renewable energy sources, such as solar, wind, and geothermal. This transition is not only driven by environmental concerns but also by technological advancements and shifting regulatory landscapes. As a result, energy companies must be able to adapt quickly to new market demands, technological innovations, and regulatory changes [2].

Principles of agile leadership

The core principles of agile leadership include transparency, collaboration, customer-centricity, and a focus on delivering value. These principles are particularly relevant in the energy transition for several reasons:

Agile leadership in the energy sector necessitates a foundation of transparency, which is pivotal for navigating the complexities of the energy transition and digital transformation. Transparency involves fostering a culture of openness and clear communication, ensuring that all stakeholders employees, customers, regulators, and the community are well informed and engaged in the process. This openness builds trust, which is essential for achieving the collective buy-in needed to drive significant change [3].

Transparency and communication

Transparency and communication are crucial for building trust, a key component of effective leadership. When leaders are open about their decisions, strategies, and the challenges they face, they cultivate an environment of mutual trust. This is particularly important in the energy sector, where the stakes are high, and the impact of decisions can be far-reaching. It is also essential for engaging employees, who are critical to the success of any transition. Agile leaders ensure that their teams are well informed about the company's vision, strategy, and progress. This involves regular updates, open forums for discussion, and accessible channels for feedback. Keeping employees in the loop instills a sense of ownership and commitment, motivating their teams to contribute actively to the transition efforts.

Transparency is not just about sharing information; it is about nurturing a culture where open communication is the norm. This culture is vital for the smooth implementation of changes and cooperation across the organization.

Agile leaders encourage open dialogue, where team members feel comfortable sharing their ideas, concerns, and feedback. This openness can lead to better problem-solving, more innovative solutions, and a more cohesive team effort.

In the context of the energy sector, clear and consistent communication about the energy transition and digital initiatives can alleviate uncertainties and build a unified direction. Leaders should leverage various communication channels, such as town hall meetings, digital platforms, and regular newsletters, to keep everyone informed and engaged. This multifaceted approach ensures that the message reaches all levels of the organization, fostering a sense of inclusion and shared purpose. Effective communication also involves active listening. Leaders must be attentive to the feedback and concerns of their teams and stakeholders. This two-way communication helps identify potential issues early and allows for collaborative problem-solving.

Transparency and communication are foundational to agile leadership, particularly in the energy sector's dynamic landscape. An open, communicative environment can build trust with employees, engage their teams, and drive successful, sustainable transformations.

Collaboration

The energy transition with its complex and interconnected landscape involves multiple stakeholders, from engineers and scientists to policymakers and consumers. Agile leadership emphasizes the importance of collaboration across these diverse groups, promoting cross-functional teams and partnerships. Collaboration is vital. Agile leaders understand that no single entity can address these challenges alone. They encourage partnerships between private companies, government bodies, academic institutions, and nongovernmental organizations, creating a synergistic approach to problem-solving. Cross-functional teams are a key element of collaborative leadership. Bringing together individuals from different disciplines and departments enables leaders to build an environment where diverse ideas and skills are leveraged to tackle complex issues. The diversity of thought and expertise within these teams can lead to more comprehensive and innovative solutions than those generated in siloed environments.

Furthermore, collaboration extends beyond internal teams to include external partnerships. For example, energy companies may collaborate with technology firms, academia, institutes, and with government agencies to address their business focus areas. Partnerships can accelerate the development and deployment of new technologies, streamline regulatory processes, and enhance the overall effectiveness of the energy transition.

Effective collaboration also involves transparent and open communication among all stakeholders. Agile leaders facilitate regular meetings, workshops, and forums where ideas can be exchanged, and feedback can be gathered. This open dialogue helps ensure that everyone is aligned with the shared goals and can contribute their unique insights and expertise. It also builds trust among stakeholders, which is crucial for long-term cooperation and success.

Customer-centricity

Understanding and addressing the needs of consumers is vital. Agile leaders prioritize customer-centric approaches, ensuring that new solutions are not only effective but also meet the expectations and demands of end-users. This involves developing user-friendly technologies and creating incentive programs that encourage the adoption of best practices.

As the energy sector evolves, the shift toward a more customer-centric approach is important in the adoption of new technologies and sustainable practices. Agile leaders recognize that the success of any transition depends not only on technological advancements and regulatory support but also on the acceptance and engagement of consumers. By putting consumers at the heart of their strategies, agile leaders can drive meaningful change that resonates with the public and meets their needs.

Agile leaders place a strong emphasis on customer engagement and communication. They recognize that building trust and maintaining open lines of communication with consumers are crucial for developing long-term relationships. This involves actively listening to customer feedback, addressing concerns promptly, and being transparent about the company's goals and progress. Engaging with consumers through various channels, such as social media, events, and customer service hotlines, allows agile leaders to create a more responsive and customer-focused organization.

Focus on delivering value

Agile leadership in the energy sector places a strong emphasis on delivering tangible value both quickly and iteratively. This approach involves implementing pilot projects, gathering feedback, and continuously improving based on real-world performance. Breaking down large initiatives into manageable parts is an important aspect of agile leadership, which can ensure that each step adds measurable value and addresses immediate needs.

In practice, this means initiating small-scale projects that can be swiftly deployed and tested. These pilot projects serve as a proving ground for new technologies and methodologies, allowing leaders to assess their effectiveness and make necessary adjustments before broader implementation. Through this iterative process, agile leaders can identify what works, what doesn't, and how to optimize performance. Gathering feedback is a crucial component of this approach. Agile leaders actively seek input from all relevant stakeholders, including team members, customers, and partners. This feedback loop ensures that the project remains aligned with the needs and expectations of those it serves, and it provides valuable insights for future iterations.

Focusing on delivering incremental value helps build momentum, as stakeholders can see the positive impact of each phase and are more likely to support ongoing efforts. This strategy helps secure ongoing support from stakeholders, including investors, regulatory bodies, and the community. Each successful iteration reinforces confidence in the overall direction and management of the transition. It shows a commitment to delivering real-world results and adapting strategies based on actual performance data.

Ultimately, the focus on delivering value through agile leadership drives continuous improvement and innovation. It creates a dynamic environment where projects evolve and improve in response to real-world challenges and opportunities. This approach ensures that the energy sector can adapt to rapid changes and maintain its trajectory toward a sustainable future.

Ørsted's transformation: A case study in agile leadership

A prime example of agile leadership driving successful transformation in the energy sector is Ørsted, a Danish multinational power company. Ørsted underwent a significant transformation under the agile leadership of its executives, demonstrating how flexible, adaptive strategies can propel a company to the forefront of the renewable energy industry.

Under agile leadership, Ørsted's executives set a clear and ambitious vision to become a global leader in renewable energy. This strategic vision was communicated transparently to all stakeholders, fostering a shared understanding and commitment across the organization. The leadership's flexibility in their approach allowed Ørsted to pivot and adapt as new opportunities and challenges emerged. This adaptive strategy was crucial in navigating the uncertainties of the energy transition, enabling Ørsted to stay ahead of industry trends and regulatory changes.

Ørsted promoted a culture of cross-functional collaboration, breaking down the silos that previously hindered innovation and efficiency, encouraging its departments to work together, and leverage diverse expertise from different areas, accelerating the development and deployment of its offshore wind projects. This collaborative approach enhanced problem-solving capabilities and ensured that innovative ideas from all levels of the organization could be effectively integrated into project planning and execution.

A key aspect of Ørsted's agile approach was its focus on developing customer-centric renewable energy solutions, prioritizing the needs of consumers and businesses enabled Ørsted to design and deliver projects that provided significant economic and social benefits to communities. This customer-centric approach ensured that the renewable energy projects were environmentally sustainable, also economically viable and socially responsible, which supported a stronger community spirit and stakeholder engagement.

Ørsted adopted an agile mindset of iterative improvement, implementing pilot projects and actively gathering feedback to continuously enhance its renewable energy technologies. This iterative process allowed the company to refine its strategies, address challenges promptly, and scale successful projects quickly, embracing a cycle of continuous learning and adaptation.

Agile leadership had a positive impact on Ørsted's successful transition from a traditional fossil fuel company to a leading renewable energy provider. The company's ability to adapt and innovate significantly reduced its carbon footprint and positioned it as a key player in the global energy transition. Ørsted's journey highlights the effectiveness of agile leadership in driving strategic transformation, which embraced innovation.

Agile leadership can be a powerful catalyst for change and is a key attribute of the new age of leadership needed to circumnavigate the energy transition and lead the workforce transformation. Embracing a strategic vision, fostering collaboration, prioritizing customer needs, and committing to continuous improvement can have a clear benefit in navigating the complexities of the energy transition but also set a benchmark for others in the industry [4].

Leadership in the digital age of innovation

Earlier chapters of this book have laid the foundation for understanding the importance of innovation and continuous improvement in the energy transition. Chapter 4 highlighted the significant advances in digital, Chapter 6 introduced the concept of centers of excellence to sharpen the saw of the organization, and Chapter 9 emphasized the new roles needed to succeed in the ET digital era discussing the significance of a workforce proficient in digital technologies and data-driven decision-making.

Leaders must draw on these concepts to build a culture that supports innovation and continuous improvement, creating and that encourages creative thinking, collaboration, and ongoing learning. Leaders play a crucial role in cultivating this culture by encouraging experimentation, providing resources, promoting collaboration, and recognizing achievements.

The global energy landscape is evolving rapidly, driven by the urgent need to combat climate change, enhance energy security, and promote sustainability. This shift toward renewable energy sources and the integration of digital technologies demands a transformative approach to leadership, one that places innovation at its core. Traditional energy companies are at a crossroads, where old business practices reliant on fossil fuels and established methods are no longer viable. The energy transition, marked by the adoption of cleaner, more sustainable solutions, presents multifaceted challenges that require leaders to think creatively and take bold actions. Embracing new technologies, encouraging and exploring novel business models, and creating an environment that drives continuous innovation are now essential for survival and growth in the energy sector.

Creating an innovative culture within an organization starts with leadership. Leaders in the energy sector must model the behaviors and attitudes they wish to see in their teams, such as openness to new ideas, willingness to take risks, and a commitment to continuous improvement. A culture of innovation encourages employees at all levels to experiment, fail fast, and learn from their mistakes. Encouraging collaboration is critical, as innovation often happens at the intersections of different disciplines. Promoting cross-functional teams and encouraging collaboration both within and outside the organization can spark new ideas and approaches. For example, energy companies can partner with technology firms to develop advanced analytics and AI solutions to optimize their operations, be it energy production, or project services. Providing resources is also crucial. Leaders must ensure that their teams have access to the technologies, training programs, and research opportunities. This might include setting up innovation labs, funding pilot projects, or providing grants for employees to pursue innovative ideas. Recognizing and rewarding innovation sustains a culture of innovation by acknowledging those who contribute to it through formal recognition programs, financial incentives, or career advancement opportunities.

Digital technologies are at the forefront of the energy transition. From AI and ML to advanced data analytics and the Internet of Things (IoT), these tools are transforming the way energy is produced, distributed, and consumed. Leaders must understand the potential of these technologies and be willing to invest in the and potentially integrate them into their operations. AI and ML for instance, enable predictive maintenance, optimize energy usage, and improve decision-making processes. AI can predict equipment failures before they occur, allowing for timely maintenance and reduced downtime. ML algorithms can analyze vast amounts of data to identify patterns and optimize energy distribution, enhancing efficiency and reducing costs. Leaders must champion the adoption of AI and ML by investing in technology, building in-house expertise through training programs and hiring skilled professionals, and ensuring that AI solutions are seamlessly integrated into existing operations and processes.

Advanced data analytics plays a pivotal role as data is the new oil of the energy sector. By leveraging big data, energy companies can make informed decisions that enhance performance and drive innovation. To harness the power of data analytics, leaders should promote data literacy, ensuring that employees at all levels understand the value of data and how to use it effectively. Implementing robust data management systems for collecting, storing, and analyzing data securely and efficiently is important because it acts as an enabler for data-driven decision-making where decisions are based on data insights rather than intuition.

Despite the clear benefits of innovation, several barriers must be overcome by leaders. Cultural resistance is a significant challenge, as employees accustomed to traditional ways of working may hesitate to adopt new technologies or approaches.

Regulatory challenges also pose a significant hurdle. The energy sector is heavily regulated, and new technologies and business models often face substantial regulatory hurdles. Leaders may engage with regulators and policymakers to advocate for frameworks that support innovation while ensuring safety and compliance. Financial constraints are another barrier, as innovative projects often require significant upfront investment. Leaders must balance the need for innovation with financial prudence, seeking partnerships, grants, and other funding sources to support their initiatives.

Innovation requires a strategic vision that aligns with the broader goals of sustainability and digital transformation. Leaders must articulate a clear vision for the future, one that inspires and motivates their teams to embrace innovation. This vision should encompass ambitious sustainability goals and embracing digital transformation as a core component of business strategy. Collaboration and partnerships with technology providers, research institutions, and other stakeholders are essential to drive innovation and achieve common goals.

The role of remote and hybrid work

The COVID-19 pandemic has precipitated a significant shift in how organizations operate, with remote and hybrid work models becoming the new norm across various industries, including the energy sector. This transition was initially driven by necessity as lockdowns and social distancing measures forced companies to adopt remote work practices. However, the benefits of these models have become increasingly apparent, leading many organizations to embrace remote and hybrid work as a permanent fixture.

Remote work has demonstrated that many tasks can be performed effectively outside the traditional office environment. This realization has prompted energy companies to rethink their operational strategies, balancing the need for on-site presence in certain roles with the flexibility offered by remote work for others. Hybrid models, which combine remote and on-site work, provide a balanced approach that can enhance productivity, employee satisfaction, and overall operational resilience [5].

Remote teams

Effective leadership in a remote or hybrid work environment requires a different set of skills and strategies compared to traditional in-office management. Leaders must adapt to the unique challenges posed by physical distance, including maintaining team cohesion, ensuring clear communication, and fostering a sense of belonging among remote employees.

One critical aspect of managing remote teams is establishing clear expectations. Leaders should communicate goals, deadlines, and performance metrics transparently, ensuring that all team members understand their responsibilities and how their work contributes to the broader organizational objectives. Regular check-ins and progress reviews can help keep everyone aligned and on track.

Additionally, leaders must leverage technology to facilitate collaboration and communication. Tools such as video conferencing (Camera On Please!), project management software, and instant messaging platforms are essential for maintaining connectivity and enabling real-time interaction among team members. Adopting these tools can help leaders to maintain the engagement and productivity of their employees and track progress of key tasks.

The importance of communication, trust, and flexibility in remote leadership

In a remote or hybrid work environment, effective communication is paramount. Leaders must prioritize open and transparent communication channels to keep team members informed and engaged. This involves not only disseminating information but also actively listening to employees' concerns, feedback, and ideas. Regular team meetings, virtual town halls, and one-on-one check-ins can help foster a culture of open dialogue and ensure that employees feel heard and valued.

Trust is another critical component of successful remote leadership. Leaders must demonstrate trust in their employees' ability to manage their time and complete tasks independently. Micromanagement can erode trust and hinder productivity, so it is essential for leaders to focus on outcomes rather than processes. Empowering employees to take ownership of their work can cultivate a sense of accountability and autonomy. Flexibility is also vital in managing remote teams. Recognizing that employees may face different challenges and circumstances while working remotely, leaders should adopt a flexible approach to work hours and expectations. This flexibility can help accommodate individual needs and promote a healthy work–life balance, ultimately contributing to higher job satisfaction and retention.

Several energy companies have successfully implemented remote and hybrid work models, providing valuable insights into effective leadership in this context. For instance, Shell has adopted a hybrid work model, allowing employees to split their time between the office and remote locations. Shell's leadership has focused on creating a flexible and supportive work environment, emphasizing the importance of clear communication and trust. The company has also invested in digital tools and platforms to facilitate seamless collaboration and ensure that remote employees remain connected and engaged.

The shift to remote and hybrid work models requires energy sector leaders to adopt new strategies and approaches. Effective remote leadership hinges on clear communication, trust, and flexibility. Adopting these principles and leveraging digital tools enable leaders to successfully manage remote teams, ensuring that their organizations remain agile, productive, and resilient in the face of ongoing change.

Leaders need to embrace digital literacy

As the energy sector is undergoing a significant transformation the integration of digital technologies has become paramount. One striking example of this shift is the appointment of Chief Artificial Intelligence Officers (CAIOs) across 22 Dubai government entities, including the Dubai Ministry. This move highlights the critical role that agile and innovative leadership will play in navigating the complexities of the new digital era amidst the energy transition.

CAIOs are a new role that has evolved as part of the new age of digital and the prolific growth of AI technology. They at the forefront of leveraging AI to drive organizational transformation. In Dubai, the appointment of CAIOs reflects a strategic commitment to embedding AI into the fabric of government operations. These leaders are tasked with developing and implementing AI strategies that enhance efficiency, improve service delivery, and foster innovation. Their roles are crucial in ensuring that AI technologies are not only adopted but also effectively integrated into existing systems and processes.

For the energy sector, the role of CAIOs highlights the broader need for leaders who can harness the power of digital technologies. These leaders must be adept at understanding and applying AI, machine learning, and data analytics to optimize operations and achieve sustainability goals. The appointment of CAIOs in Dubai serves as a model for how other sectors, including energy, can benefit from dedicated leadership focused forward thinking and digital transformation.

The role of CAIOs in Dubai is a prime example of how embracing innovation can lead to transformative outcomes. These leaders are responsible for identifying and implementing cutting-edge AI solutions that can revolutionize government services. Similarly, energy sector leaders must be willing to explore and invest in new technologies, such as AI and automation, to optimize energy production, distribution, and consumption [6].

The future of leadership in the energy sector

As the energy sector continues to evolve, the role of leaders will become increasingly complex and dynamic. The appointment of CAIOs in Dubai illustrates the importance of having dedicated leaders who can navigate the digital transformation and drive innovation. Energy sector leaders must similarly embrace agility and innovation, prioritizing the integration of advanced technologies and fostering a culture of continuous improvement. The future of leadership will be defined by the ability to adapt, innovate, and leverage digital technologies. Drawing inspiration from initiatives like the appointment of CAIOs in Dubai, energy leaders can position their organizations for success in the new digital era and the ongoing energy transition.

Energy Leaders must become aware of the benefits in using digital tools and AI technologies. This involves more than just adopting new software or platforms; it requires a fundamental shift in how leaders think about and engage with technology.

Firstly, leaders need to recognize that data is a strategic asset and be well-versed in data and data analytics. Understanding how to collect, analyze, and interpret data is crucial for making evidence-based decisions. This skill set enables leaders to identify trends, anticipate challenges, and seize opportunities more effectively. Furthermore, leaders must foster a data–driven culture within their organizations, encouraging all employees to utilize data in their daily operations.

Secondly, leaders must leverage AI to enhance operational efficiency and innovation. AI can automate routine tasks, freeing up employees to focus on more strategic and creative endeavors. Leaders who harness the power of AI can drive significant improvements in performance and competitiveness.

Leaders must also stay abreast of emerging technologies and continuously seek to integrate them into their business strategies. This requires a commitment to ongoing learning and a willingness to experiment with new tools. Staying ahead of the technological curve will support leaders to that their organizations remain agile and resilient in a rapidly changing environment.

Examples of digital and AI-driven leadership strategies

Several energy companies have successfully integrated digital and AI–driven leadership strategies, serving as exemplars for others navigating the digital transformation.

One notable example is Shell, which has embraced digital transformation through its AI and data analytics initiatives. Shell has developed a comprehensive digital strategy that leverages AI and big data to optimize its oil and gas operations. This transformation was spearheaded by leaders who understood the potential of digital tools and prioritized their integration into the company's core operations.

Equinor has been at the forefront of using AI to drive innovation and business transformation. Equinor's AI initiatives have been utilized across various operations, from offshore drilling to renewable energy projects, providing intelligent insights and automating complex processes. Equinor's leadership has been instrumental in fostering a culture of innovation, encouraging employees to experiment with AI and develop new applications for the technology.

Leadership in the digital era requires a profound shift in mindset and approach. Leaders must embrace digital transformation and AI, leveraging these technologies to enhance decision-making, drive efficiency, and foster innovation. By understanding and utilizing digital tools and AI, leaders can position their organizations for success in an increasingly digital world.

What does a leader of the future look like?

As the global energy sector undergoes monumental transformations, the traditional leadership styles that have dominated for decades are increasingly called into question. The rapid evolution of the energy landscape necessitates a new breed of leader those who prioritize innovation, sustainability, and agility over conventional metrics of performance. Historically, CEOs focused on numbers and bottom-line impacts, often resorting to downsizing and cutting human labor costs to balance the books. However, in an era where the energy sector must pivot toward greener portfolios and embrace digital transformations, such approaches are increasingly seen as outdated. The leaders of the future must navigate new challenges and uncertainties, requiring them to think strategically, break the mold, and make courageous decisions regarding project delivery models.

Agile leadership emerges as a vital paradigm for the energy transition and digital era. Agile leaders prioritize adaptability, empowering their teams to pivot strategies in response to new information and emerging trends. This approach is particularly relevant in the context of the digital revolution, where advancements in AI, machine learning, and data analytics continuously reshape industry operations. The successful implementation of these technologies hinges on leaders who can guide their teams through ongoing changes, fostering a culture of learning and experimentation.

Agile leadership involves being responsive to new information, technologies, and market conditions, making decisions that align with evolving goals. Agile leaders are characterized by their openness to innovation, willingness to take calculated risks, and ability to pivot when necessary. This agility allows leaders to stay ahead of the curve, continuously adapting to the fast-paced changes defining the energy transition and digital era.

Digital technologies are at the forefront of the energy transition. From AI and ML to advanced data analytics and the IoT, these tools are transforming how energy is produced, distributed, and consumed. Leaders must understand the potential of these technologies and be willing to invest in and integrate them into their operations. AI can predict equipment failures before they occur, allowing for timely maintenance and reduced downtime. Machine learning algorithms can analyze vast amounts of data to identify patterns and optimize energy distribution, enhancing efficiency and reducing costs. Leaders must champion the adoption of AI and machine learning by investing in technology, building in-house expertise through training programs and hiring skilled professionals, and ensuring that AI solutions are seamlessly integrated into existing operations and processes.

Future leaders in the energy sector must be adept at leading workforce transformation to navigate the complexities of the energy transition. This involves several key strategies:

Leaders must foster a culture of continuous learning within their organizations. This includes providing ongoing training and development opportunities for employees to acquire new skills and knowledge. With the rapid advancement of digital technologies, it is crucial for the workforce to stay updated on the latest trends and innovations. Leaders can implement mentorship programs, partnerships with educational institutions, and encourage participation in industry conferences and seminars to ensure their teams remain at the forefront of technological and industry developments.

In a sector characterized by rapid change, agility and adaptability are essential traits for the workforce. Leaders should encourage a flexible approach to work, where teams are empowered to pivot and adjust strategies as needed. This involves fostering an environment where experimentation and calculated risk-taking are valued. By promoting a mindset that embraces change and innovation, leaders can help their teams navigate the uncertainties of the energy transition more effectively.

The integration of digital technologies, such as AI, machine learning, and advanced analytics, is crucial for optimizing operations and driving efficiency. Leaders must champion the adoption of these technologies, ensuring their teams are proficient in using digital tools to enhance performance. This may involve investing in new software, training programs, and hiring experts in digital technologies.

Collaboration across different functions and disciplines is vital for addressing the complex challenges of the energy transition. Leaders should promote the formation of cross-functional teams that bring together diverse expertise and perspectives. This collaborative approach can lead to more innovative solutions and accelerate the implementation of new technologies and practices. Effective collaboration also extends beyond the organization to include partnerships with external stakeholders, such as technology providers, research institutions, and regulatory bodies.

As the energy sector shifts toward more sustainable practices, leaders must prioritize sustainability in all aspects of the business. This involves setting clear sustainability goals, tracking progress, and engaging employees in sustainability initiatives. Leaders can create awareness campaigns, provide sustainability training, and integrate sustainable practices into everyday operations.

Effective communication is essential for leading workforce transformation. Leaders must ensure that their vision, strategy, and progress are clearly communicated to all employees. This involves regular updates, open forums for discussion, and providing platforms for employees to share their ideas and feedback. Transparency in decision-making processes and being open about the challenges and successes of the transition can build trust and motivate employees to engage actively in transformation efforts.

A diverse and inclusive workforce brings a wealth of perspectives and ideas that can drive innovation and problem-solving. Leaders should prioritize diversity and inclusion initiatives, ensuring that all employees feel valued and have equal opportunities for growth and development. Creating an inclusive work environment allows leaders can harness the full potential of their workforce and foster a culture of collaboration and respect.

The leaders of the future in the energy sector will be those who can adapt, innovate, and lead with a vision that embraces both the challenges and opportunities of the energy transition and digital era. They will be transparent communicators, collaborative partners, customer-centric strategists, and agile innovators, driving their organizations toward a more sustainable and technologically advanced future.

References

[1] R.A. Heifetz, Transformational Leadership and Change Management, Harvard University Press, 1996.
[2] H. Van Swaay, Agile Leadership: A Leader's Guide to Orchestrating Agile Transformation, iUniverse, 2021.
[3] S. Hayward, The Agile Leader: How to Create an Agile Business in the Digital Age, Kogan Page, 2018.
[4] Orsted Company Website, Leading the Transformation to Green Energy, 2022.
[5] T. Neeley, Remote and Hybrid Work Models Post-COVID-19, Harper Business, 2021.
[6] The National News, UAE, Dubai Appoints 22 Chief AI Officers to Boost High-Tech, 2022.

Index

Note: Page numbers followed by *f* indicate figures and *t* indicate tables.